ALPHA GOD

ALPHA GOD

The Psychology of Religious Violence and Oppression

Hector A. Garcia

59 John Glenn Drive
Amherst, New York 14228

Published 2015 by Prometheus Books

Alpha God: The Psychology of Religious Violence and Oppression. Copyright © 2015 by Hector A. Garcia. All rights reserved. No part of this publication may be reproduced, stored in a retrieval system, or transmitted in any form or by any means, digital, electronic, mechanical, photocopying, recording, or otherwise, or conveyed via the Internet or a website without prior written permission of the publisher, except in the case of brief quotations embodied in critical articles and reviews.

Cover design by Nicole Sommer-Lecht

Prometheus Books recognizes the following registered trademarks mentioned within the text: Axe®, Burger King®, Rolex®, DSM-5®, Humvee®.

Inquiries should be addressed to
Prometheus Books
59 John Glenn Drive
Amherst, New York 14228
VOICE: 716–691–0133
FAX: 716–691–0137
WWW.PROMETHEUSBOOKS.COM

19 18 17 16 15   5 4 3 2 1

Library of Congress Cataloging-in-Publication Data

Garcia, Hector A., 1970-
 Alpha God : the psychology of religious violence and oppression / by Hector A. Garcia.
     pages cm
 Includes bibliographical references and index.
 ISBN 978-1-63388-020-7 (pbk.) — ISBN 978-1-63388-021-4 (ebook)
 1. God. 2. Violence—Religious aspects. 3. Psychology, Religious. 4. Dominance (Psychology) 5. Sexism in religion. 6. Evolutionary psychology. I. Title.

BL473.G355 2015
202'.113—dc23

2014039221

Printed in the United States of America

For my parents

CONTENTS

CHAPTER 1: ENTER GOD THE DOMINANT APE	11
Dominance Defined	15
History: How a Dominant Male God Rises to Power	18
CHAPTER 2: EVOLUTIONARY MECHANISMS: ETIOLOGY	27
Natural Selection	28
Sexual Selection	29
Mate Competition	29
Mate Selection	30
Mating Strategies	32
Quantity	32
Male Jealousy	34
Quality	36
Kin Selection and Kin Altruism	37
Evolutionary Psychology and the Science of God	39
CHAPTER 3: THE PROTECTOR GOD	45
Protector Males	45
Paternal Certainty in Apes, Men, and God	52
Problems of Divine Alliance Making	57

CHAPTER 4: SEXUAL DOMINANCE: FROM APES TO MEN TO GODS 63

Apes   63
- Violence and Sexual Access   63
- Infanticide in Nonhuman Primates   66

Men   68
- What Men Want   68
- What Dominant Men Get   69

Gods   73
- The Lustful Godhead   73
- Sexually Repressive Gods: Divine Jealousy   74
- The Virgin and the King   76
- Chaste and Submissive   79

Women   83
- What Women Want in Their Men and Gods   83
- The Cost to Women and Children   86
  - Veiling   87
  - Violence against Women   88
  - Infanticide in Men and God   90

A Case Study   92

CHAPTER 5: COOPERATIVE KILLING, IN-GROUP IDENTITY, AND GOD   99

Evidence in the Microcosm   99

Establishing Boundaries with Kin Altruism   101
- In-Group, Out-Group   101

Reciprocal Altruism and Indirect Reciprocity   106

God as War-Maker   109
- Patterns of Primate Alliance-Making   109
- Costly Signals with God for Help in Killing   111

The Great Out-Group Prejudice of Humankind   117

The Sociopathy of the In-Group	120
Sociopathic Killing	123

CHAPTER 6: WHAT IT MEANS TO KNEEL — 131

Size and Domination: What it Means to Be Big	131
Big Heads, Big Hats	132
Posturing	136
Eye Contact	140
Hand and Foot Kissing	141
Submission by Ideological Surrender	144

CHAPTER 7: MALADAPTIVE SUBMISSION TO THE GODHEAD — 151

The Pecking Order	151
Worthlessness and the Sin of Pride	153
Anhedonia	159
Sex and the Sin of Lust	159
Food and the Sin of Gluttony	165
Diminished Ability to Think	169

CHAPTER 8: THE FEARSOME REPUTATIONS OF APES, MEN, AND GODS — 177

The Origins of Reputation	177
Men	180
Gods	186

CHAPTER 9: GOD'S TERRITORY — 195

Marking Territory — 196
Territory: Staking Claim to Sex — 202
Rape and the Bible — 207
Staking Claim to Mother Earth — 211
    The Earth as Ecosystem — 212
    Male Competition and Resource Consumption — 215
    Religious Rapacity: An Alternate View — 218

CHAPTER 10: RIGHTING OURSELVES — 223

The Psychology of the Other — 224
Pacifism and Selective Observance — 230
Erecting a Wall — 236
Societal Health and Future Directions — 243
Closing Thoughts — 248

ACKNOWLEDGMENTS — 251

NOTES — 253

INDEX — 283

Chapter 1

ENTER GOD THE DOMINANT APE

> If oxen and horses and lions had hands and were able to draw with their hands and do the same things as men, horses would draw the shapes of gods to look like horses and oxen to look like oxen, and each would have made the gods' bodies to have the same shapes as they themselves had. —Xenophanes (ca. 570–ca. 478 BCE)

What is God? Many would say that God is love, or God is beauty. For others, God is an immaterial being, the creator of the universe. God has been described as compassionate and merciful, as the ultimate moral authority, or the ultimate source of goodness in the world. The pious draw from this vision of God a sense of awe, purpose, hope, and empathy. From this vision, masses of people around the world convene around a shared sense of wonder, appreciation, and unity, and they cultivate between one another an environment of kindness, generosity, and support. This vision of God is indisputable, insofar as it forms the phenomenology of the religious worship of God.

But there is another vision of God that is just as real. The majority of the world's believers worship a god that is fearsome and male, and his portrayal demands reckoning. Scripture depicts this god as one who rains fury upon his enemies and slaughters the unfaithful. It also shows him policing the sex lives of his subordinates and obsessing over sexual fidelity. Extremists, drunk on this vision, steer airplanes into buildings or obliterate themselves in crowded marketplaces. They foment sexual shame and engage in genital mutilation, acid attacks, and so-called honor killings. They start inquisitions and witch hunts, religious wars and religious conquests. They seize ideological control and breed superstition, ignorance, and prejudice. And they also seek to enforce a prohibition against questioning God, leaving such inhumanities unexamined, sometimes for fear of the treatments just described.

Critically, we now live in an age in which religions clash with women's rights as gender equality strains against its margins, in which theocratic regimes are gaining control of nuclear arms, and in which dangerous fundamentalism is increasingly taking hold around the world. This is a crucial moment for us to force the wedge of inquiry, if only to better understand the means by which religion may be used to encourage what is worst—rather than what is best—in human nature.

We may begin by questioning whether there is something common to the perpetrators of the kinds of violence and oppression listed above. The seemingly obvious answer is that these acts are almost exclusively committed by men. In the rare instances where they are committed by women or children, the acts are almost always influenced or coerced by men. This is an important starting point. Since another common root to these acts is purported religiosity, a second key question becomes, is there something common to the vision of God behind them? Here we arrive at the crux of the matter: the common vision is that of God *as* man.

I argue here that God was created in the image of man. This argument is not new. The epigraph of this chapter would suggest that thinkers have made this connection since at least the time of the ancient Greeks. However, there are good reasons not only to emphasize that God was created in the image of man, rather than the other way around, but also to study the dominance characteristics portrayed in God—most notably because men of power have historically conflated themselves with God in order to secure more power and have used this power to enact further violence and oppression. This pattern has emerged again and again across religious history as men have summoned divine legitimacy to justify their worst impulses. God himself is frequently portrayed as engaging in violent acts, thus serving to validate the destructive actions of the powerful.

This is most evident among the Abrahamic religions (Judaism, Christianity, and Islam), whose scriptures all too frequently depict a despotic male god. The Abrahamic god is the most widely worshipped *man-based* god, with followers comprising over 50 percent of the world's religious practitioners.[1] And certainly a tyrannical God is not unique to the Abrahamic traditions; many (even polytheistic) religions around the world have dominant male gods who go around behaving like dominant male humans. I will thus occasionally reference male gods from other religions to illustrate how consistently male-typical patterns of dominance traverse traditions of faith. Even so, the Abrahamic god will remain my focus here, if only because He is by far the most globally dominant.

To understand such a god, we must first understand the minds of men, for it is these minds that think up ways to oppress and kill. Arguably the best way to understand the ultimate basis for male violence and oppression is through the evolutionary sciences. Such disciplines reveal the ancient underlying motivations for violence and oppression, molded as they were by the process of natural selection. The patterns of behavior such motivations were passed on by our primate ancestors and are easily evidenced in living nonhuman primates—our closest living relatives. Despite his upright stance, his clothes, and his sometimes-good table manners, man rarely surpasses his most primal impulses. Accordingly, men often seek out dominance in the manner of male apes, using violence to obtain evolutionary rewards such as food, territory, and sex. Humankind may have managed to create things like tools, weapons, and religions, but we remain one species of great ape that emigrated from Africa.

This can be difficult for many people to hear because as humans we have a tendency to think of ourselves as unique and to hold ourselves above other species. But we have DNA like all other life-forms—which ultimately shapes or brains and influences the manner in which we think—and we share as much as 99 percent of our DNA with nonhuman primates. And like other animals, we are organic beings that live, eat, reproduce, and die. As such, we require things like food, sex, and territory to fulfill our organismic destinies. None of this should be surprising.

However, it *may* be surprising to realize that while the god of the Abrahamic religions has powers that humans do not, He remains unnecessarily preoccupied with what are ultimately very human, and very ape-like, concerns. God is portrayed as being omnipotent (possessing infinite power), omniscient (having infinite knowledge), omnipresent (present everywhere), immaterial (without bodily form), and eternal, meaning that he never dies. This raises the question—Why should such a God concern himself with such pedestrian pursuits as food, sex, and territory? Why demand food as a sacrificial offering or order the conquest of biblical lands if he has no need of either to survive and can create worlds by simply speaking them into being?

The answer is that God is an alpha male, a dominant ape. In other words, depictions of the Abrahamic god, and of male gods from religions around the world, reflect the essential concerns of our primate evolutionary past—namely securing and maintaining power, and using that power to exercise control over material and reproductive resources. Understanding God therefore requires an understanding of man's evolved legacy within primate social hierarchies; and understanding religious violence and oppression requires taking a careful look at how thoroughly we have projected our own psychology onto our vision of the sacred.

This book examines how God has been drawn in human form, complete with an ancient repertoire of behaviors inherited from our primate male ancestors. It examines how male primates struggle for dominance within social groups, using a variety of strategies—fear and aggression among them—to acquire rank status. Rank, in turn, typically confers rewards, which for males includes preferential access to resources such as food, females, and territory. Dominant apes and men have a long history of securing such biological treasure by perpetrating violence and oppression on lower-ranking members of their societies. Once we observe that God, too, is portrayed as having great interest in these kinds of resources, and as securing them through similar means, it becomes increasingly clear that He has emerged as neither more nor less than the highest-ranking male of all.

Thus this book aims to illuminate patterns of dominance behaviors in God, tracing them back to their origins in men, and illustrating them in extant nonhuman male primates—all to show how humans have created gods that are intuitive to their evolved psychology, and with such devastating consequences. To accomplish this I call upon scientific research in the fields of evolutionary biology and psychology, clinical psychology, primatology, and world history, as well as theoretical formulations that have yet to be tested empirically. It is a matter of no mean importance that we come to better understand our gods, for only in doing so can we hope to understand the role they play in rationalizing human violence and brutality.

It is important to point out that I neither make, nor intend, any overarching attack on religion. Rather, I make the argument that our evolutionary drives have limited the reach for goodness in religions because they—like our religions—evolved in a savage world where survival was tenuous, and where aggression promoted survival. Human potential is so vast, but we may have limited ourselves by the gods we created.

For those who may take offense at the very premise of this book, it may also be worth mentioning that I am not claiming that your god really is a dominant ape. In fact, I am arguing that in reality there is no supernatural being, or any kind of superordinate consciousness, out there that resembles men or apes, in neither form nor behavior. My own opinion is that if there is a higher power (and I have yet to see evidence that there is) it certainly doesn't resemble any of the obvious, simplistic, and species-centric characterizations that have been widely proposed throughout the history of religion. Rather, as I will attempt to show, such characterizations arise from our evolved psychology, which is strongly dedicated toward navigating interactions with other humans, particularly those with power—an ancient task that remains critical for our present-day survival. If I am right, then the deviant sides of God really implicate our own conceptual limitations and have little to bear on that "higher power," however defined.

I understand that the kinds of questions that I raise in this book can run the risk of being taken out of context and used, just as religion has been used, to justify out-group hatred, or violence, even worse; and this gives me pause. However, better understanding religious violence and oppression is so crucial to curbing the human suffering it causes, that the risk of asking provocative questions must be taken. To this end, we require a better understanding of what instinctive drives we, as creatures of biology, bring to religious belief and practice, for it is these drives that are ultimately behind every form of violence and oppression.

DOMINANCE DEFINED

In order to take a more evolutionarily informed look at God, it is worth taking a bit of time to clarify a few terms and basic evolutionary ideas. First of all, what is dominance? And what does it mean to primates such as humans?

All great-ape species have male dominance hierarchies[2] and there is a relative lack of female dominance hierarchies among the great apes.[3] Males, with the notable exception of bonobos, typically dominate females.⁴ Dominance status is often associated with greater male violence; dominant chimpanzees, for example, show agonistic displays more often, start aggressive interactions more often, escalate aggression more often, and win aggressive interactions more often than their lower-ranking counterparts.⁵

Humans, like other primates, for the greater part live in hierarchical societies. While the degree to which human societies are rank-stratified varies across cultures, there is rank structure even in relatively more egalitarian hunter-gatherer societies.⁶ Rank has important implications for behavior.

Anthropologists Joseph Henrich and Francisco Gil-White emphasize that high status involves a system of rewards in which males receive "preferential reproductive access to females, food, and spaces, as well as a disproportionate amount of grooming from others" (privileges of high status that will be explored throughout this book).⁷ The authors also note an important characteristic of human hierarchies, which is that rank status may be maintained through either "*dominance* (force or force threat)" or through "*prestige* (freely conferred deference)."⁸ In dominance hierarchies, status is reinforced with aggression and fear, and the lower-ranking typically avert eye contact, yield space, groom their superiors, and make other submissive gestures. Prestige, as described by the authors, is characterized by the relative absence of fear and is maintained by high-ranking individuals demonstrating merit, skill, wisdom, or persuasiveness. Rather than averting eye contact and seeking greater distance, lower-ranking individuals seek eye contact and proximity with prestigious individuals—often in order to gain valued information. In illustrating the difference between dominance and prestige, the authors evocatively offer the great paraplegic physicist Stephen Hawking as the exemplar of pure prestige, and the high-school bully as the exemplar of pure dominance.

In this book I will focus on dominance. As noted by Henrich and Gil-White, individuals may use both dominance and prestige to achieve and maintain status. With this in mind, I will infrequently reference prestige in male status as it occurs in men and in God, with the understanding that dominant individuals may vary their techniques to achieve status goals.

Even so, the majority of the book will narrow in on dominance behaviors in male apes, men, and the Abrahamic god, because He typically follows patterns of dominance rather than prestige, as described by Henrich and Gil-White. One can argue that the Abrahamic god, and perhaps particularly the figure of Jesus Christ in the Christian New Testament, also makes use of prestige as a strategy for achieving status; however, the Abrahamic god's use of dominance is robust, and the use of fear to maintain rank is widely documented. Since it is this use of dominance rather than prestige that most influences violence within these traditions, dominance takes center stage here. In particular, I will focus on four main components:

1. Intimidation—Dominant males use dominance displays to intimate greater size, whereas lower-ranking members demonstrate submission by intimating smaller size (e.g., shrinking down), averting eyes, or communicating emotions such as fear and humility (as opposed to anger and pride).
2. Territorial acquisition—Dominant males often control territory, which I define not only as tracts of earth but also as the control of resources within specific geographic boundaries. These resources have key implications for males' evolutionary fitness, most importantly food and females, both of which dominant males commandeer upon winning territory.
3. Sexual control—Following evolutionary drives, dominant males often monopolize sexual access to females. They will mate-guard and show great rage and sexual jealousy when their sexual claims are challenged. They often spend great energy attempting to stave off the sexual ambitions of their male rivals.
4. Violence—Dominant males will enact violence to establish and maintain rank status, and the resources associated with it. Sometimes this involves killing.

To begin to understand how such appetite-driven tactics might have become associated with our notion of the divine, it helps to understand how combative much of human history has been. Perhaps nowhere is that history more poignant than in the Middle East during the biblical age, when the turbulent forces of humanity were forging the identity of the Abrahamic god that we have today. In the history outlined below we may also begin to understand what the meteoric spread of a dominant male god owes to his intuitive appeal, particularly for primates whose minds, by way of biological evolution, come predisposed to fear, to submit to, and to follow dominant males.

HISTORY: HOW A DOMINANT MALE GOD RISES TO POWER

History reveals in striking form how men have historically conflated themselves with God as a means to amplify power, and how male gods rise to totalitarian rule in the manner of men—through violence and killing. With a critical read of history, we are also able to account for the Abrahamic god's domineering temperament, which he appears to have inherited from effective warlords of the biblical age.

A number of scholars,[9] perhaps most notably Robert Wright,[10] have traced the evolution of god concepts and religious practices as humans moved from hunter-gather societies to chiefdoms and eventually to nation-states. They illuminate an intriguing history, within which flows a rather-complex confluence of cultural, political, militaristic, and psychological forces, all shaping the countenance of God (or gods). While my main focus is on the evolutionary angle, I am obliged to give some attention to this history, for it illustrates the thunderous path toward monotheism in the Middle East that brought us the Abrahamic religions and does much to shed light on the evolving need for dominant male gods. In doing so, I favor the structure and interpretation of this history as told by Robert Wright.

Wright begins his account by outlining five categories of supernatural beings seen consistently across groups of hunter-gatherers, designed largely to explain the natural world:

1. elemental spirits (e.g., inanimate phenomena, such as wind, moon, stars, with personality and soul);
2. natural phenomena controlled by supernatural beings (e.g., a personified deity who controls the wind);
3. organic spirits (e.g., coyote spirits, tree spirits);
4. ancestral spirits; and, importantly,
5. high gods (which Wright describes as "a god that is in some vague sense more important than other supernatural beings and is often a creator god").[11]

All of these classes involve human projections. Now, because we evolved in ranked societies, the natural extension is that our projections are ranked socially. Even in early religions we can observe male dominance behaviors in conceptions of god, particularly "high gods." For example, the Native American Klamath tribe's sun god, Kmukamtch, was jealous of his son, Aishish, and spent a great deal of energy trying to seduce Aishish's wives.[12] Gaona, the dominant god of the !Kung San of Africa, raped his son's wife and ate two of his brothers-in-law.[13] Both sexual acquisitiveness and mate competition are dominance behaviors common among male primates, which we will explore in later chapters.

Historical scholars, including Wright, argue that as people moved from nomadic bands of hunter-gatherers into larger agrarian societies, their religious needs changed to reflect a new lifestyle.[14] With larger populations, the roles of gods began to reflect social concerns rather than the forces of nature. Notably, gods began to more actively regulate social interactions and punish breeches of morality and cooperation. Wright argues that hunter-gatherers, who lived in small, close-knit societies, had little need for gods to oversee in-group processes; these societies were small and transparent enough to be regulated from within. As societies grew into larger chiefdoms, gods became "the guardians of political power, supervisors of economic performance, and supporters of social norms that let unprecedentedly large numbers of people live together."[15] For example, the Tongan gods of Polynesia were said to punish theft with shark attacks.[16] In chiefdoms, dominant men often possessed divine authority, or were walking gods on earth. Breaking the chief's laws became the same as breaking the laws of God.

As populations grew to state levels, the power of the gods grew commensurately with the power needed to regulate greater masses of people. Similarly, the tasks of the gods became more specialized, reflecting the growing complexity of social order and greater division of labor. As states began to subsume ethnically diverse bands of people, there was a need not only for regulating in-group behavior but also for regulating interactions between states. As diverse peoples and their gods came into contact, political leaders began to create rules reflecting international law, steeped always in religious belief—in this era, religion, law, and politics were inextricable. In order to foster trade and commerce, and simply through sustained contact, many states either incorporated the gods of their trading partners or at the very least tolerated them. Thus many early religions in the Middle East remained polytheistic. But even here, under relatively peaceable coexistence with other gods, dominant male gods set to the male-typical business of establishing territory.

International-relations theorist Adam Watson writes, for example, that in third millennium BCE Mesopotamia, "Enlil, the king of all the lands and the father of all the gods, marked out a boundary for the god of Lagash and the god of Umma by his decree. The king of Kish measured it out in accordance with the word of the god of legal settlements, and erected a stone boundary marker there."[17] The territoriality of male gods began to take on greater proportions, following the evolutionary dictates of the powerful men who represented them, a line that was often intentionally obfuscated. For example, when the king of Umma violated Enlil's decree, he was "punished by the army of Lagash (or as the historical record has it, by the god of Lagash through the army of Lagash)."[18] And so the needs of gods began to transform to include transnational territories and codes of behavior for larger and larger masses of people.

Gods moved toward monotheism alongside high-ranking men who were increasingly able to consolidate entire populations under autocratic rule. One example comes from Babylon, an important enemy of ancient Israel, where the Abrahamic god was born. Hammurabi, a Babylonian king in the early second millennium, championed the god Marduk and utilized Marduk's image to materialize his own political ambitions. Hammurabi, credited with creating one of the first legal codes, was sure to emphasize that his law-making capacities were divinely authorized by the gods Anu and Enlil. The two gods, according to Hammurabi's code, then promoted Marduk to supremacy and assigned him "dominion over earthly man."[19] Marduk's image was steeped in references to mate-competition—he was a glorious, sexually potent, dominant male god (described as having the heart of a kettle drum and the penis of a snake that produced golden sperm, for example).[20] With the lines blurred between god and man, more specifically between Hammurabi and Marduk's supremacy, the way was greased for Hammurabi's dominion of all Mesopotamia. Though Hammurabi died before this was realized, Babylon ultimately managed to defeat Mesopotamia, and Marduk became the head of the pantheon of gods across the subsumed territories, either by subordinating the resident gods of the conquered or by sequestering their functions. For instance, Adad, who was once the god of rain, became the "Marduk of rain." Nabu, who was once the god of accounting, changed into "Marduk of accounting."[21] But still, this was an example of *monolatry*, rather than monotheism—meaning that one god ruled over other gods in the pantheon, rather than that one god was the only god in all existence. Like a pantheon of bureaucrats, there were many other gods during this period of ancient Babylon, each performing different functions; in the ninth century BCE, a census turned up around sixty-five thousand gods.[22]

With the identities of male leaders and male gods so intertwined, one could understand the motivation to consolidate divine identities—doing so could prove a fast-track to consolidating power, whereas leaving a host of other gods in position had the potential to confuse (and ultimately dilute) the king's power structure. But again, it should be recognized that dominant male gods are the creations of men (rather than women) specifically. The bellicose ancient cradle of civilization was not a cradle of gender equality. Powerful men wrote the doctrine, designed the laws, and set the special relationships with their gods. The historical uses for these gods were the male-typical pursuits of territory and power, following an ancient legacy of male primate behavior. The meshing of gods and kings may have also dissuaded rebellion, or as historian Will Durant remarked, "All the glamor of the supernatural hedged about the throne, and made rebellion a colossal impiety which risked not only the neck but the soul."[23]

Other gods have assumed *monolatristic*, if not virtual monotheistic, positions. One example comes from Egypt, another important enemy of ancient Israel. The Egyptian god Amun came to dominate the pantheon of Egyptian gods, a position gained by a series of successful Egyptian military campaigns that Amun symbolically spearheaded. Amun became known as the "king of gods," the "prince of princes." He became an ultimate god, the greatest of transcendent deities of Egyptian religious history, and his priests amassed great political and economic power from riches won in military campaigns. As Wright tells it, when Pharaoh Amenhotep IV inherited the throne upon the death of his father, he may have had good reasons to feel threatened by a god with such power.[24] It wasn't long before Amenhotep deposed Amun and put Aten (a sun god) in his place, and then declared himself Aten's son. Aten eventually became the creator of the world. In the story of Amenhotep and Aten we can see not only human power and territoriality exercised through the gods, but also human jealousy. Under Amenhotep, any person named Amun was forced to change his or her name. Similarly, any depiction of Amun was erased from existence, from wherever it appeared. Though Aten symbolically embodied both the masculine and feminine, this level of possessiveness is traceable to the behaviors of despotic men. Joseph Stalin, for example, used similar tactics when he assumed control of Russia and had the images of men fallen from favor erased from history books, currency, and political sculptures; he had most of these men exterminated altogether. It is worth noting that this business of marking over the territorial markers of one's rivals is known to primates and proliferated among the gods of the biblical age.

The ancient Israelites, too, were at first polytheistic. Yahweh was a god that emerged from a pantheon of other gods and was initially neither monotheistic nor possessing of many of the transcendent qualities he later came to embody. There are many references to other gods in the Bible, and in the history of Israel, before monotheism took hold.

It is important to understand the geopolitical environment of the ancient Middle East when the first Abrahamic religion, Judaism, and its monotheistic god, Yahweh, were gestating. Conquest involving the slaughter of entire cities was not uncommon. This was violence conducted up close and personal, where you made eye contact with your attacker before he physically hacked you to pieces (with dull bronze and later iron weapons), killed your entire family, and burned your city to the ground. Unless you possessed great power, or allied yourself with someone who did, you were subject to raids or conquests of these kinds. Lesser states often paid tribute to more powerful kingdoms for protection lest they be annihilated. Powerful men appointed male gods as their generals, and success in battle (or in genocide) was often attributed to the gods. Of course, this is not the only epoch of human history in which agonies of this kind occurred, but as great populations began to emerge in the cradle of civilization, there was a corresponding growth of the scale of warfare.[25]

It is not difficult to understand how, in environments marred by warfare and existential uncertainty, the religions of a people would reflect a need for a fearsome, protector male god. Likely these gods reflected the warlord kings of those epochs, who performed the same function for their people—men who rose up and took arms against invading enemies, or who invaded enemies for their resources. Further, ancient Israel was at a distinct disadvantage because it was situated between two superpowers of its time, Assyria and Egypt, and was often subject to slavery and slaughter between the two. Forming alliances with one or the other was not much of an option, as Wright describes it, because for "a small state wedged between two great powers, 'alliance' often amounts to vassalage."[26]

Moreover, kings in this region were not only taxed with aggressive encounters from outside groups—they were also required to maintain a stable in-group hierarchy, which often necessitated aggression. Sectarian groups within a nation's borders could weaken the entire state, rendering it less effective at coordinating common goals such as defense. More often than not, these fissions were headed by their own dominant male leaders (and their godheads), who could handicap the force of a nation with petty squabbling. But fiercely dominant male leaders had the power to punish factionalism and to cement alliances. Israel's strategic and size disadvantages made the need for in-group cohesion all the more urgent. As always, the men who shouldered the task of fusing in-group schisms did so at the level of humans and gods simultaneously.

Within a history of blood-loss and slavery, a series of powerful kings in Israel began addressing the need for a more powerful god. In around 640 BCE, King Josiah assumed the throne of Israel and began to consolidate power under Yahweh, simultaneously removing traces of other gods. Deities of the region—for example, Astarte, Chemosh, Milcom, and Baal—mostly syncretic gods formed from centuries of cultural merging, were deemed abominations.[27] All altars, idols, and other religious accoutrements of the gods were destroyed, along with their temple priests. According to the Bible, Josiah "slaughtered on the altars all the priests of the high places who were there, and burned bones on them" (2 Kings 23:20). Scripture began to police dissent as means to ensure consolidation. Of Josiah, the Bible reads, "Anyone who does not heed the words that the prophet shall speak in my name, I myself will hold accountable. But any prophet who speaks in the name of other gods, or presumes to speak in my name a word that I have not yet commanded that prophet to speak, that prophet shall die" (Deut. 18:19–20). Similarly, if anyone dared to suggest worshipping other gods, that person should be killed, "even if it is your brother, your father's son, your own son or daughter, or the wife you embrace, or your most intimate friend" (Deut. 13:6–9). And if you came across an Israeli town worshipping other gods, "you must certainly put to the sword all who live in that town. Destroy it completely, both its people and its livestock. Gather all the plunder of the town into the middle of the public square and completely burn the town and all its plunder as a whole burnt offering to the LORD your God. It is to remain a ruin forever, never to be rebuilt" (Deut. 13:15–16).

This strategy of intolerance was a design of Josiah's ambitions to expand the Israeli empire, first by uniting southern and northern Israel under one god. Though Josiah's prescriptions for nonbelievers were brutal, such was the tenor of this merciless age. For instance, later when King Zedekiah of Judah rebelled against the Babylonians, King Nebuchadnezzar captured Jerusalem, burned it to the ground, and killed Zedekiah's son in front of him, gouged out Zedekiah's eyes, and took the greater part of the city into Babylon as slaves. Knowing the importance of the godhead (perhaps even believing it), Nebuchadnezzar was sure to destroy Yahweh's temple.

Israel, like other states of the region (for instance, when Israel subjugated the Moabites), also suffered through many years of conquest and slavery—it served, for example, King Cushan-Rishathaim of Aram-Naharaim for eight years, and King Eglon of Moab for eighteen years. The Israelites were also exiled and held in captivity for nearly fifty years after being conquered by the Babylonians during the fifth century BCE (a conquest the Babylonians saw as evidence of the supremacy of their god Marduk). Wright goes so far as to argue that this stint in exile was the most profound in Judaic history and an important catalyst in the development of monotheism—for just as national religion can unite a people, so can national trauma.[28] From their trials and humiliations, the Israelites began transforming God to a redeemer, and eventually solidified Yahweh's singular divinity. The concept began evolving from roughly, "You shall have no other gods before me" to "There are no other gods."

The Israelites began developing what has been described as *retribution theology*. From this age of suffering comes a litany of redresses that Yahweh begins to extract from the oppressors of Israel, in the fashion of dominant human lords of the ancient Middle East. Yahweh proclaims in the book of Isaiah, "I will make your oppressors eat their own flesh, and they shall be drunk with their own blood as with wine. Then all flesh shall know that I am the LORD your Savior, and your Redeemer, the Mighty One of Jacob" (49:26). For Israel's enemies, Yahweh promises fearsome retribution:

> Because you have clapped your hands and stamped your feet, rejoicing with all the malice of your heart against the land of Israel, therefore I will stretch out my hand against you and give you as plunder to the nations. I will cut you off from the nations and exterminate you from the countries. I will destroy you, and you will know that I am the LORD. (Ezek. 25:6–7)

The list of retributive threatening goes on and exemplifies the need for a warrior god to protect and avenge His people against their subjugators. In a political environment run by strongmen, one needs a god based on strongman psychology, one equipped for maneuvering within the dominance hierarchies of men. My intention in reviewing this history is to show the context in which the three great monotheistic religions arose. Notably they arose from an explosion of population density in the Middle East, where merciless slaughter at the hands of men encouraged the people to turn to strong, warlike leaders (to protect against the other men of a similar bent), and to gods who were similarly fearsome.

Many of the patterns outlined in this history have occurred before, repeating for millions of years across the savage, primeval landscapes of our primate ancestors. As in ancient Mesopotamia, the savannas and rainforests of Africa have witnessed dominant males leading bloody incursions, capturing territories, killing in-group males, and protecting against outsider males assembled ominously at the border. Humans are unique in having created supernatural agents that perform these functions alongside their mortal counterparts. The religio-cultural contexts in which dominant male gods arise are rooted in our biological heritage. In the next chapter we come to understand how natural selection shaped the human mind, influenced the litany of dominance behaviors described in the history above, and engendered the cross-cultural tendency to create supernatural agents who demand our allegiance and watch over us while we sleep.

Chapter 2

EVOLUTIONARY MECHANISMS: ETIOLOGY

In this chapter I will recount explanations of how humans evolved to anthropomorphize the natural world and how this tendency ultimately resulted in man-based gods. For this the basics of evolutionary science is required. I will also elaborate on evolved sex differences in reproductive strategy. This understanding is crucial because mate competition is the ultimate driving force behind male violence. The broader point of this chapter, however, is that in order to understand a man-based god, we must first deconstruct its model. This will properly place God's despotism into the primate context from which it descended.

We begin with the epoch-making contributions of Charles Darwin. From the decline of the Roman Empire up until the Renaissance, Europe was marred by cultural, economic, and intellectual deterioration. These Dark Ages were a time of wide superstition and ignorance, thanks in great part to religious authorities who worked vigorously to control the minds of the people. Fast-forward a thousand years to the Enlightenment, when a growing momentum began to free the world's base of knowledge from the dark sepulchers of religious belief where it had been stagnating for centuries. Teachings by scholars such as William Paley began pulling away from scripture through fields like *natural theology*. In natural theology, knowledge was acquired through common, everyday observations of the world, in contrast with the prevailing dogma of *revealed religion*—knowledge disclosed only through divine revelation. But the process of establishing scholarly independence from religion was tentative, while the distance put between church and intellectual freedom was politically cautious. Even in natural theology, all the flora and fauna of the natural world ultimately owed their existence to a creator god.

In 1859, after much deliberation, Charles Darwin unveiled his book,

On the Origin of Species. It was here that he argued that life on earth was the result of natural laws. Most important, he revealed his discovery that life's diversity was a product of *natural selection.* Natural selection is the theory that differences in traits lead to differences in the ability to survive; in turn, traits better suited for survival are passed on to subsequent generations.

Here I wish to highlight two things. The first is that the scope of Darwin's contribution was profound. This book, along with numerous scientific fields, would be impossible without *On the Origin of Species.* The second is that Darwin's discoveries had momentous implications for human identity. His theory of natural selection paved the way for a godless evolutionism in which individual species were not placed here separately, but in which all life's vast diversity has a common origin. Moreover, it forced people to contemplate their place in the universe, which has since generated both great existential anxiety and great intellectual liberation.

NATURAL SELECTION

Natural selection is an algorithmic process by which nature impels genetic design across generations. Genes are swirling strands of protein chains floating around the nuclei of cells that themselves do not think or choose but blindly operate by the algorithm. The two necessary components for natural selection are *individual differences* and *differential reproduction.* In short, each one of us houses different arrangements of genes, and, across generations, these differences are what allow our species to adapt to a changing environment. The genes that produce traits suited for survival on average confer a reproductive advantage (fitness) and therefore get expressed more often than other genes. Nature doesn't really care which genes are reproduced, and the natural traits we possess aren't necessarily nice, moral, fair, or kind. Sometimes they are, but that is because those particular expressions of genes produce an algorithmic advantage. It all boils down to individual differences and differential reproduction.

SEXUAL SELECTION

Mate Competition

In addition to natural selection, Darwin pointed out that sexual selection also drives evolution. There are largely two forms of sexual selection; one is *mate competition*. In species that compete for the right to mate, individual differences conferring a competitive advantage are selected for and therefore shape the course of evolution. Bull elk, for example, have giant antlers—and hulking neck and back muscles, among other features—which they use to compete for mates during the rutting season. These traits were inherited from their ancestors; those among them that possessed larger racks on average won more antler-wrestling competitions, gained more access to females, and were more successful at reproduction. Contrarily, scrawny elk with smaller antlers were less equipped for competition, reproduced less, and ultimately possessed traits that became evolutionary dead ends. Boar tusks, curled ram horns, and flowing lion manes (which make them look bigger), are all products of mate competition.

Male aggression (including in humans) is another trait shaped by mate competition.[1] This connection is supported by a large body of evidence from diverse scientific fields, including anthropology, which shows differential reproduction in action. For example, Yąnomamö tribesmen of the Amazon who have killed men in battle have more wives and more children than those who have not killed.[2] Evolutionary psychologist David Buss explains the many fitness benefits of aggression, which "include expropriating resources, defending against incursions, establishing encroachment-deterring reputations, inflicting costs on rivals, ascending dominance hierarchies, dissuading partner defection, eliminating fitness draining offspring, and obtaining new mates."[3] This list does much to illuminate the aggressive behaviors of mortal men who must compete for resources with other men if they wish to survive. However, we will see in later chapters how God applies aggression to the same survival tasks as men, violating the central notion of omnipotence.

Mate Selection

Mate selection is the other main force in sexual selection. Generally, males and females are programmed for selecting successful traits in their partners. In other words, they are genetically predisposed to look for qualities in their prospective mates that are likely to maintain or improve the survival of their offspring. Some traits were developed for courtship displays; male peacocks, for example, bear giant, shimmering, and iridescently colored tail plumes. On the surface, such ostentation may seem like evolutionary fat strangely uncut by the economizing dictates of natural selection. However, the ability to maintain costly displays like a peacock tail is a sign of health and success, implying not only that the cock was successful at acquiring the necessary resources to grow it but also that he has managed to escape predators even while pulling his weighty, flamboyant plume.[4] Further, if peahens choose cocks with the most impressive tail fans, then they are likely to produce impressive male offspring that appeal to the lady-peacocks—this is known as the *sexy son* hypothesis.[5] In turn, sexy sons produce more copies of their mother's genes.

Similarly, elk cows, like females of other species, are not passively waiting for a victor whom they receive with no say. Females typically prefer the victors of rigorous competition because the victors often bear traits of strength, courage, and determination, which benefit their offspring. Females preferring weak and demure males would produce weak and demure male offspring, which would not win mate competitions or survive other challenges of the natural world—meaning weak-male preference would quickly disappear from the gene pool. This is one reason why females of many species, including women, are hostile toward the advances of lesser males but attracted to dominant males.

Another appeal of dominant males is that they protect females from other males, as is evident among other primates. Primatologist Frans de Waal notes of chimpanzees that "a female who is being threatened may run to the most dominant male and sit down beside or behind him, whereupon the attacker will not dare to proceed."[6] American primatologist Barbara Smuts found that among savanna baboons, females befriend males who protect them (and their offspring) from other males, often in exchange for sex.[7] Smuts and David Buss both argue that men protecting women from other men has been a critical facet of humankind's evolutionary past. Buss even says that given the high incidence of rape in many cultures, the ability to provide protection remains important to female mate selection in the modern day.[8] This may explain why women worldwide tend to prefer taller men—because in this world, bigger generally means stronger. This preference is borne out fairly conclusively across studies.

For example, researchers in one study took height information of 720 couples and found that only one couple was a pair in which the woman was taller than the man, a ratio far rarer than the chance probability of the woman being taller, which was 2:100.[9] Another study found that height is related to evolutionary success, with taller men being more likely to have children than shorter men.[10] More directly testifying to female attraction, research has also found that women tend to prefer taller men when they are at the most fertile phase of their menstrual cycle.[11] In short, in a world of often-brutal physicality, size usually confers dominance, which women (on average) prefer. The relevance of male dominance to female reproductive success offers great insight into the relationships women have with dominant, male, protective gods—a topic to which we shall return.

Mate selection also applies to male preference. One example is men's fascination with women's physical features. In a world where childbirth is medically risky (and child rearing is physically demanding), physical health equates to reproductive viability. Long before medical technology provided insight to a female's reproductive capacity, physical appearance was the best cue. Another example is men's attraction to youth, a preference that has also been shaped by health, as well as by the comparatively narrow span of female fertility. First, women's reproductive capacity is limited to their youth, a relatively brief period spanning from adolescence to menopause. Second, the younger a woman is within that span, the more children she potentially can birth and raise over time. And third, on average, younger women are physically healthier than older women and thus are more suited for the high physical demands of reproduction. In sum, youthful, healthful-looking women are seen as attractive because genes in men coding for attraction to these traits had a survival advantage over time.

MATING STRATEGIES

Quantity

Male apes, men, and man-based gods show a tendency to monopolize females. It is therefore important to understand where man-based gods acquired this habit. Like many animals, men and women generally employ different mating strategies. In using the term *strategy* I am not saying that men and women have sex with the literal intention of winning child-birthing contests; they don't. As psychologist Steven Pinker puts it, men and women have sex because it feels good, and it is the "strategy" of our genes to make it feel good as a means to replicate themselves into future generations.[12]

The strategic differences roughly boil down to quantity versus quality. For women, the overarching strategy is quality. Ova are costly to produce and are limited; a woman's fertility—the availability of an ovum—is limited to monthly ovulation and is often diminished by breast-feeding. Further, women have the lifetime capacity to produce about five hundred ova *total.* In short, it pays for women to be selective about with whom they choose to share this precious resource.

By contrast, in men, the overarching strategy is quantity. Sperm are cheap, produced by the millions, and available between the vast span of adolescence and old age. Theoretically, this means that the most productive male strategy is to reproduce with as many females as possible. In practice, this means that men who like novelty in sexual partners and who are driven by their lust to seek out variety (e.g., through casual affairs, numerous sexual partners across the lifespan, etc.) are going to pass on their genes in higher numbers; genes coding for sexual acquisitiveness have an algorithmic advantage over those genes, say, coding for sexual *contentment.* This trait does not particularly favor notions of gender equality, but it's simple, efficient, and incredibly easy to see in men around the world.

Research uncovers this male strategy from many angles. The sexual refractory period, normally referring to males, is a period of recovery after intercourse—after ejaculation when the male is unable to achieve an erection. Research shows that this period quickly ends with the presence of a novel female.[13] Once a novel female is introduced, the male gets an erection and copulation is on—and on and on, continuing with each novel female. One study of rhesus monkeys found that male arousal continued to decline with familiar females even when the females were made constantly aroused by hormone injections.[14] The decline quickly reversed when novel females were introduced, suggesting that male arousal is less tied to female arousal than it is to being an evolutionary "quantity" strategy employed by males.

This is certainly not the strategy that is always employed. Buss reminds us that casual sex, multiple partners, or extramarital affairs can be risky; they can expose a male to venereal diseases or a husband's shotgun, or they can cost a great deal of time and energy.[15] But compared to women, casual sex for men is relatively cost-free. A man theoretically can walk away from casual sex with scarcely an ounce more investment than a deposit of cheap sperm. For a woman it's different. You never hear women brag, "I don't have any children . . . *that I know of.*" They always know because they have borne the burden.

David Buss's lab has produced a wealth of information about human mating strategies In one study, men were found to prefer more sexual partners than women, to need significantly less time than women to consent to sex, and to have lower standards than women on the characteristics of the partners of a casual affair. Further, out of sixty-seven characteristics of potentially desirable traits in the mate of a casual affair, men had lower standards than women for forty-one of them. In *The Evolution of Desire*, Buss's seminal work on the evolutionary strategies of human mating, he points out that

> men require lower levels of such assets as charm, athleticism, education, generosity, honesty, independence, kindness, intellectuality, loyalty, sense of humor, sociability, wealth, responsibility, spontaneity, cooperativeness, and emotional stability.[16]

The depths to which men's standards will sink are pretty staggering, but they are not surprising when viewed through an evolutionary lens. When subjects were asked about undesirable characteristics, men had fewer qualms about traits such as "mental abuse, violence, bisexuality, dislike by others, excessive drinking, ignorance, lack of education, possessiveness, promiscuity, selfishness, lack of humor, and lack of sensuality." In contrast, men in this sample rated only four traits as significantly less desirable than did women: "low sex drive, physical unattractiveness, need for commitment, and hairiness,"[17] all of which speak to problems in fertility, or, as in the case of commitment, a relational impediment to the numbers strategy.

Male Jealousy

Man-based gods show great sexual jealousy, which I will expose later. For now, it is important to understand where male gods acquired this emotion. One might think that the worst evolutionary outcome in the grand competition to survive and reproduce is *not* reproducing. However, there is something far worse—investing precious time and resources in an unrelated individual's offspring, or what is known as *cuckoldry* (named after the cuckoo bird that suckers other birds into tending its eggs and raising its young). Rearing young requires the provision of vital resources—resources that were especially risky to obtain in the harsh landscapes of our ancestors. It also requires protection from predators and the violence of other males. Notably, caring for offspring is more costly the longer offspring stay dependent. While most primates pay high costs for rearing offspring, humans bankroll the most. And even though women have historically been the biggest providers of childcare, and men have a long history of walking away from offspring, men have also provided for women and children. Given the evolutionary stakes, it makes sense that male sexual jealousy would have arisen through natural selection. Jealous males, on average, would have avoided cuckoldry, thus passing on genes coding for jealousy. Non-jealous males would have weeded themselves out of the gene pool by unknowingly caring for another male's offspring, a behavior that can involve decades of provisioning.

This is not to say that women don't also experience jealousy—but there are important sex differences in the manner in which the lurid green monster surfaces. While men tend to be more jealous of women's sexual infidelity, women tend to be more jealous of men giving resources, love, and affection to other women. Experimentally, this has been an intensely robust finding across cultures,[18] and is rooted in reproduction; women require that love and resources stay with them and their children and do not get shared with their competitors, while men require the certainty that the children they are supporting are not their competitors'.

Male gods often require chastity, and the threat of cuckoldry explains where they acquired this concern—men. Sex differences in the value of chastity are borne out empirically as Buss and his colleagues discovered in a study on mate preferences of an enormous and highly diverse international sample of 10,047 individuals.[19] Men in the study consistently placed higher value on spousal chastity than did women. Further, women tend to go out of their way to market themselves to the evolutionary imperative of men by signaling chastity. Some women, for example, will even undergo painful and expensive surgery to repair their ruptured hymens—the thin vaginal membrane that signals, albeit somewhat unreliably, that a women is a virgin. Men rarely concern themselves with proving their virginity or even their sexual modesty. However, men have shown themselves to be immensely concerned with those traits in females and have created numerous customs—from virgin brides to chastity belts to burkas—in order to ensure success in their evolutionary strategies.

Male gods have also been known to cloister women, yet another behavior they inherited from primates (more on this in later chapters). Male primates of many species will guard fertile females. Generally, the more dominant the male, the more successful he is at this enterprise. Men do this as well. Moreover, research finds that men can sense, largely unconsciously, when their partner is fertile. One study had women wear T-shirts at different phases of their menstrual cycle and asked men to rate the smell of those T-shirts on "pleasantness" and "sexiness."[20] Men consistently rated T-shirts worn by women during the follicular phase (when the female egg is ready for fertilization) higher on sexiness. Other studies show that men guard their mates more when they're in the fertile phase, for example, by monopolizing their time, calling them unexpectedly, or getting angry when they talk to other men.[21] With the help of male-dominated cultures (and religions based on dominant male gods), some men have been able to keep harems, thus combining mate guarding with the quantity strategy.

In sum, males are prone to aggressive mate competition, to experiencing powerful jealousy, to enforcing female chastity, to mate-guarding (sometimes behind cloistered walls), and to employing a *quantity*, or, *numbers* reproductive strategy. Later I show how God is described as employing all of these hallmark evolutionary strategies of men.

Quality

In discussing the role of dominant male gods in our lives, female strategy must also be understood. Across the enormous population of monotheists in the world, billions are women. And while women have often been coerced into following restrictive religious dictates, many follow dominant male gods with genuine enthusiasm. Much of this concerns the evolutionary relationships women have had with dominant male humans. Women are primed to be far more particular in the *quality* of their mates, and dominance can be a strong indicator of mate quality among men.

The quality strategy can best be understood by examining differences in reproductive investment across the sexes. Women devote great resources and take high risks in producing offspring. Pregnancy and childbirth is medically risky; women to this day die in childbirth, while men, obviously, never do. Also, across cultures, women are usually burdened with the preponderance of childcare. In addition, as we have discussed, ova are scarce commodities. In the aggregate, these factors strongly tip the scale of resource investment on the side of females. With all that is at stake, it pays for women to be far more selective than men when it comes to choosing a mate.

Because raising human offspring requires so much time, energy, and resources, selective pressures drive female preferences toward men who have resources and are committed to investing them in both the females and their children. When science is put to the task, female mating strategies emerge conclusively around the globe. Here Buss reports on findings from his international study: "Women across all continents, all political systems (including socialism and communism), all racial groups, all religious groups, and all systems of mating (from intense polygyny to presumptive monogamy) place more value than men on good financial prospects. Overall women value financial resources about 100 percent more than men do."[22]

Now, here might be a good place to insert an indignant argument about this being a function of unequal pay scales in the workplace. However, recall that 99 percent of our development as a species occurred outside the modern hyperindustrialized world in environments drastically different than those in which we now find ourselves—notably without the modern-day workplace. One might also argue that the disparity in earning potential between men and women impels women to value male investment. However, the questionnaire that Buss and his colleagues used was from a study of Americans conducted in 1939, then replicated in 1956, again in 1967, and again in the mid-1980s. Across decades in which there was enormous upheaval in gender norms in the workplace in America, the results were robust—women valued financial prospects about twice as much as did men. Similarly, across the world, where we would expect to find a great deal of cultural difference, women consistently preferred good financial prospects in a mate about twice as much.

Not surprisingly women tend to like older, higher-status men because these qualities often mean more resources. Again, in Buss's international studies, these preferences were consistent in women worldwide, across religions, political ideologies, and geography. While women are attracted to prestige, power, status, and high rank, Buss writes that they are cool to men who are weaker, lower in station, or likely to be dominated by other men. In short, power comes with resources, which matters greatly in a world where resources are needed to survive. This means that women in our evolutionary history who were attracted to men with power on average reproduced more successfully than women less interested in power.

KIN SELECTION AND KIN ALTRUISM

I will later show how, like humans and other animals, the Abrahamic god favors his offspring. For now we must understand kin selection, an evolutionary force that explains family favoritism. Kin selection theory, first developed in 1964 by British evolutionary biologist William D. Hamilton,[23] is the idea that biological relatives can influence each other's evolutionary fitness. This process is fairly clear when considering a parent-child relationship; parents nurture their children, and each parent shares 50 percent of his or her genes—and so parental investment ensures that his or her genetic legacy proceeds down the evolutionary stream.

Individuals who tended to recognize genetic relatives (kin recognition) and behave differently toward relatives versus nonrelatives (kin discrimination) tended on average to pass on more copies of the genes underlying these behaviors. This has never been a perfect science. Rather, in general, organisms look for heuristics to discern relatives such as similar visual appearance and smells.[24] Curtly put, "if they look like me, they are probably a relative; therefore, I should help them."

These concepts are vastly important to our understanding of evolution because they point to the gene, rather than to the individual, as being the unit of selection. Genes tend to influence behaviors that result in more copies of themselves, even when those copies will exist in other organisms. British evolutionary biologist Richard Dawkins, author of *The Selfish Gene*,[25] goes so far to say that organisms can be seen as the "survival machines" of genes. In other words, genes produce traits in organisms that best suit the propagation of themselves.

This also explains altruistic behaviors, seeming evolutionary quandaries that come at a cost to individual fitness. Consider saving from a burning building: your child, your nephew, your fourth cousin, your goldfish, and your houseplant. Chances are you would save them in that order and your willingness to risk your life would decline as you went down the list. Understanding kin altruism, then, it is not surprising that your willingness has a strong positive correlation to the amount of shared genetic material with the ones waiting behind the flames. If you happen to perish as a result of your altruistic act, by saving your child at least you pass on 50 percent of your genes, your nephew 25 percent, and so on. As British geneticist J. B. S. Haldane famously said, "I would lay down my life for two brothers or eight cousins."[26] In sum, kin altruism is another way to pass on genetic material—by helping genetically related individuals survive and reproduce. While the idea of god operating by the rules of kin discrimination and kin altruism violates the notion of an everlasting god (who has no need for biological reproduction), such notions form central tenets of the Abrahamic religions, as I will later explain.

EVOLUTIONARY PSYCHOLOGY AND THE SCIENCE OF GOD

Understanding a god that behaves as if his mind is programmed by evolution requires that we understand the evolved human mind. Evolutionary psychology—deriving from evolutionary biology, cognitive psychology, and neuroscience—is an approach to psychology that endeavors to understand the influence of ancient selection pressures on the nature and development of the human brain. Given that we spent over 99 percent of our history as hunter-gatherers, many of the mental capacities we now possess were specialized to handle problems of survival in a hunter-gatherer lifestyle.

In their primer to evolutionary psychology, Leda Cosmides and John Tooby recall a noted passage by American psychologist William James in which he rightly notes the need to "make the natural seem strange" in order for the evolutionary adaptations of human psychology to be recognized, individually examined, and placed within the context of natural selection:

> It takes . . . a mind debauched by learning to carry the process of making the natural seem strange, so far as to ask for the *why* of any instinctive human act. To the metaphysician alone can such questions occur as: Why do we smile, when pleased, and not scowl? Why are we unable to talk to a crowd as we talk to a single friend? Why does a particular maiden turn our wits so upside-down? The common man can only say, *Of course* we smile, *of course* our heart palpitates at the sight of the crowd, *of course* we love the maiden, that beautiful soul clad in that perfect form, so palpably and flagrantly made for all eternity to be loved!
>
> And so, probably, does each animal feel about the particular things it tends to do in the presence of particular objects. . . . To the lion it is the lioness which is made to be loved; to the bear, the she-bear. To the broody hen the notion would probably seem monstrous that there should be a creature in the world to whom a nestful of eggs was not the utterly fascinating and precious and never-to-be-too-much-sat-upon object which it is to her. Thus we may be sure that, however mysterious some animals' instincts may appear to us, our instincts will appear no less mysterious to them.[27]

To say "of course" in this context is to use sloppy, species-centric reasoning and to bypass explanatory information. Thus, "making the natural seem strange" is an excellent means to achieve the reflective distance required to examine unconscious psychological phenomena as they occur in the observer. This will be required throughout this book, particularly for the religious reader for whom worshipping a dominant male god feels perfectly intuitive.

Cosmides and Tooby argue that "the brain is a physical system. It functions as a computer with circuits that have evolved to generate behavior that is appropriate to environmental circumstances."[28] The point of the analogy is that the brain is an immensely complex array of interconnected neural circuitry designed to perform certain functions. The output of this is the mind. All our thoughts, hopes, dreams, intentions, fears, emotions, sense of others, sense of self—everything comprising our experiential universe—are functions of the electro-chemical functioning of the brain.

Such is also true for religious experience. One piece of evidence for the neurological basis of religious experience comes from research on temporal lobe epilepsy (TLE). During seizures in the medial temporal area of the brain, the afflicted report profoundly religious experiences in which they sense complete unity with God and the universe.[29] There are corrective surgical procedures for TLE, but few sufferers consent to them. And who would want to? Religious ecstasy feels wonderful. And this fact, perhaps better than many, points to religiosity as having benefitted the survival of our ancestors. As a general rule, what feels good to us (e.g., sex, eating) has been good for our survival, just as what feels bad to us (e.g., extreme heat or cold) has probably been bad for our survival.

It was once believed that the brain was an all-purpose problem-solving mechanism that was a blank slate from birth and that all of our mental capacities were entirely learned. This notion of a blank slate, or *tabula rasa*, was passed down into Western and Middle Eastern thought by Greek philosophers and was made popular by English philosopher John Locke in his seminal work *An Essay Concerning Human Understanding*. What we have since learned about the mind lends insight into why we see our gods as alpha males.

The evolutionary sciences have begun to reveal that rather than blank slates, we are actually possessed of an enormous array of specialized neural circuits to solve specific problems—known as modularity. Often "mental modules" show themselves when they get differentially impaired. One good example occurs in autism. Autistic children have an impaired ability to impute mental states to others—to *mind read*, as the skill is known, that is, to understand that others have thoughts, emotions, beliefs, intentions, false intentions, or everyday feats of mental acrobatics (which most people take for granted) such as, "I know that you know that I know that you know something," and so on.[30] However, research finds that mind reading does not hinge on general intelligence, indicating that it is a separate specialty. For instance, when compared to children with Down syndrome (who are mentally retarded but able to "mind read"), matched on IQs, autistic children do far worse at mind reading.[31] Mind reading, and offshoots thereof, end up being crucial functions for humans, with implications for our god projections, as scholars in the rapidly emerging science of religion, including the *evolutionary psychology of religion*, have begun to explain. Of note, some theorists in these disciplines describe religion as an adaptation that facilitated cooperation among early humans.[32] Some describe it as a by-product of existing, evolved cognitive capacities.[33] And still others describe it as both.[34]

In order to understand the projection of god as a dominant male primate, it is useful to ask how we have come to project at all, and why our projections tend to look human. Anthropologist Stewart Guthrie makes a good case for our tendency to attribute human characteristics to God having arisen from cognitive adaptations designed to detect human presence in our environment, or what he describes as "systematic anthropomorphism."[35] He starts by introducing a concept known as *animism*, or the tendency to attribute life to inanimate things. In nature, it pays to have false-positive-leaning life-detection systems (versus false-negative) because animate beings can impact survival. Per Guthrie's analogy, if I am a hiker and I see a large brown object in the distance, I might be inclined to think "bear" or "boulder." However, it pays to favor bear, even if I am wrong. If I think "bear," and take precautions, but it ends up being a boulder, then I keep hiking along and no harm is done. However, if I think, "just a boulder" and it turns out to be a grizzly, the bear could end up shredding me to pieces. This concept squares well with what we know of natural selection; those with false-negative-leaning detection systems would be food for bears more often, would reproduce less often, and therefore the genes coding for such systems would fall from the gene pool.

Guthrie describes how humans are strongly inclined to anthropomorphize because the most important animate being to human survival has historically been other humans. From a cognitive-design perspective, Guthrie is right on. That we possess distinctive neural machinery specifically for "mind reading" is only one of many pieces of evidence. In fact, we possess an enormous amount of neural architecture dedicated to detect, understand, communicate with, learn from, cooperate with, fight with, and mate with other humans. From a safety perspective, this is also clear; we don't install locks on our doors to keep out the wolves or the rattlesnakes—we do it to keep out other humans. Philosopher and religious scholar John Teehan points out that detecting other humans is not an easy task because humans can hide, use camouflage, and act from a distance using weapons and traps, and that, for these reasons, a detection system that only saw humans when they were clearly present would be a dangerous strategy.[36] Human deceptions such as these would make seeking human presence even when it is not there an important adaptation, one that likely pushed the development of an exquisitely powerful human-agency-detection capacity.

Building on Guthrie's work, psychologist Justin Barrett has referred to cognitive strategies designed to overdetect agency as Hyperactive Agency Detection Devices (HADD).[37] HADD appears to be particularly adapted to imputing agency to ambiguous stimuli. Research has demonstrated the brain's eagerness in action—for instance, in the research lab, people quickly attribute agency to simple geometric shapes moving on a screen that bear no resemblance whatsoever to humans or animals.[38]

The natural extension is that we project supernatural agency to natural phenomena and that those agents behave like humans. This is incredibly easy to observe: natural disasters are the will of God; disease is the possession of an evil spirit; good fortune is the beneficence of a generous deity; and stars are the embodiments of humanlike gods. The fact that these projections are arrived at biologically does much to explain the tendency for invisible spirits, ghosts, gods, angels, and demons to surround every human culture on earth, in one form or another. Aspects of humanity that are universally shared often reflect our shared biology.

French anthropologist Pascal Boyer[39] also describes religion as having arrived by way of ancient cognitive machinery that was already designed to make inferences about the world—such as those related to social exchanges, moral infractions, animate beings, and the like. Per Boyer, certain doctrines happen to mesh well with these existing machineries and therefore produce emotionally powerful and memorable experiences (whereas the underlying reasons for these experiences are unconscious).

While theoretical formulations about religion have been made for millennia, the empirical study of religion is in its infancy and much more research is needed. However, in terms of its origin, the alpha-god paradigm fits well with Boyer's description—we have in place cognitive capacities designed to detect social (perhaps even male) dominance. Research has found that we possess certain brain structures that specialize in processing rank information,[40] and as a result we process such information automatically and at great speed.[41] This is not surprising in a species that spent so many millions of years evolving in rank-structured groups, where understanding dominance hierarchies was critical for survival. In sum, this means that humans have mental machinery designed to light up in the presence of dominant males, that imputing dominance is largely unconscious, and that religious belief has evolved to mesh with those machineries, thus they produce powerfully intuitive experiences that reflect our evolved genetic makeup.

With this base of knowledge in evolutionary science, we now turn to how the existence and personalities of male gods betray our evolutionary projections and reference a world in which the presence of dominant males ultimately shaped the evolutionary landscapes of religion. We start with the idea of a protector god.

Chapter 3

THE PROTECTOR GOD

PROTECTOR MALES

The natural world writhes with giant snakes, crocodiles, lions, leopards, hyenas, and countless other fearsome predators. As such, death by predation was a pressure that shaped the evolution of our brains. One outcome is our enduring fear of "monsters." As philosopher David Livingstone Smith argues, this fear is exemplified in modern-day Hollywood horror movies, which captivate viewers by stimulating the predator modules of their brains—within the confines of the movie theater, where they can safely experience an ancient charge of adrenaline.[1] Smith cites environmental philosopher Paul Shepard:

> Our fear of monsters in the night probably has its origins back far in the evolution of our primate ancestors, whose tribes were pruned by horrors whose shadows continue to elicit monkey screams in dark theaters. As surely as we hear the blood in our ears, the echoes of a million midnight shrieks of monkeys, whose last sight of the world was the eyes of a panther, have their traces in our nervous system.[2]

What both Smith and Shepard are ultimately getting at is that our capacity for fear is shaped by our evolutionary past, just as in other animals. As an example (cited by Smith), pronghorn antelope once ran for their lives from cheetah tailing them across American grasslands. Today, pronghorn are the fastest land animals in North America, capable of speeds over sixty miles per hour, while the cheetahs that once hunted them have long since disappeared. Noting their gratuitous speed, biologist John A. Byers said that the pronghorn are haunted by "the ghosts of predators past."[3] Charles Darwin speculated that human fears were shaped by similar forces. When he observed his two-year-old son showing intense fear of large animals at the zoo, he wondered, "Might we not suspect that the fears of children, which are quite independent of experience, are the inherited effects of real dangers . . . during savage times?"[4]

A good example of Darwin's insight comes from clinical psychology. While people can, and do, develop phobias of just about anything, snake phobias are highly overrepresented among pathological fears, reflecting the dangers of the primordial environments in which our brains evolved.[5] This is true even in large, industrialized cities such as New York, London, or Tokyo, where the probability of being killed by snakes is practically zero—compared to, for example, New York City cab drivers, who are a real menace to public safety, although the rate of taxi phobias is also practically zero.

We might suspect that in addition to the impulse to move away from the dangers of our evolutionary past, we have also inherited the impulse to move toward that which provides protection. I argue that such impulses are disproportionately directed toward powerful male figures, a trend that reflects a legacy of male protectors in our evolutionary past that is reanimated with the male saviors of religious doctrine.

Evidence of this legacy can be seen in living primates—across species, protection from nature's killing machines is often the specialty of males. Males tend to be bigger than females and to have longer, sharper canine teeth, which make them good at security, a duty they often assume. For instance, when traveling to feeding grounds, male baboons patrol the periphery of the troop to ward off or attack predators, while females and infants travel in the protected center. More plainly, male patas monkeys have been directly observed attacking jackals, and male langur monkeys, attacking raptors, each in defense of infants.[6] Male baboons[7] and chimps[8] will also ward off and kill leopards. Similarly, it is usually the men in hunter-gatherer societies who kill dangerous predators such as cheetahs, leopards, lions,[9] and pythons[10] that pose a threat to their communities.

Because the primeval social environments in which our brains evolved required us to seek protection from powerful males against dangerous predators, such a role in male gods is emotionally intuitive. The natural extension is that gods now protect us from predation.

The ancient Egyptian god Bes is one example. Bes's primary job was to protect women and children, and he was known for his ability to strangle lions, bears, and snakes with his bare hands.[11] Protector male gods are seen in Greek mythology as well. Zeus, for instance, conquered Typhon, a serpent with one hundred dragon heads that had been destroying Greek cities. Likewise, Zeus's son Herakles fought with many dangerous creatures and, as a baby, saved his brother by strangling two snakes. Notably, during the famed "Twelve Labors," Herakles killed a menagerie of monsters that had been tormenting the Greeks: the Nemean lion, the Lernaean hydra, the Stymphalian birds, and other mythic predators. For his acts of bravery, Heracles was regarded a hero and fabled to have made the world safe for humankind.[12]

The Judeo-Christian male god is also tasked with protection, described variously throughout the Bible as a shield, a rock, a stronghold, a fortress, a strong tower, and a hiding place. He, too, has assumed the ancient task of dispensing with dangerous animals:

> But you, LORD, do not be far from me. You are my strength; come quickly to help me. Deliver me from the sword, my precious life from the power of the dogs. Rescue me from the mouth of the lions; save me from the horns of the wild oxen. (Ps. 22:19–21)

> You will trample upon lions and cobras; you will crush fierce lions and serpents under your feet. (Ps. 91:13)

One predator described in the Bible is actually a composite of predators found on the African savanna, including a bear (an animal once native to Africa but now extinct from overhunting):

> And I saw a beast coming out of the sea. It had ten horns and seven heads, with ten crowns on its horns, and on each head a blasphemous name. The beast I saw resembled a leopard, but had feet like those of a bear and a mouth like that of a lion. (Rev. 13:1–3)

This beast persecuted Christians until another powerful male, Jesus Christ, rode in on a white horse, vanquished it, and threw it into a lake of fire (Rev. 19:19–21).

Similarly, Satan (God's nemesis in the Christian tradition), has been associated with the serpent—that enduringly fearsome creature that tempted Eve in the Garden of Eden. He is described as a dragon in the book of Revelations: "The great dragon was hurled down—that ancient serpent called the devil, or Satan" (12:9). Revelations also has another powerful male religious figure, an angel, defeating this predator personified as Satan:

> And I saw an angel coming down out of heaven, having the key to the Abyss and holding in his hand a great chain. He seized the dragon, that ancient serpent, who is the devil, or Satan, and bound him for a thousand years. He threw him into the Abyss, and locked and sealed it over him. (Rev. 20:1–3)

When Satan is portrayed in Christian iconography, it is often in the form of a humanoid with animal body parts such as large canine teeth, horns, cloven hooves, and bat-like wings, thus evoking the image of a dangerous animal. Conceivably, the mythic antagonist to a dominant male god could take a less animalistic form, such as a person or an abstract concept like sin or temptation or immorality. But dangerous animals resonate deeply with our evolved emotions, as do the powerful males to which we are drawn when faced with mortal danger. Psychological research demonstrates that this draw is for the greater part unconscious, which suggests its evolved design.

For instance, there is a body of scientific research based on terror management theory (TMT)—the idea that humans have erected psychological defenses against the terror of death. Humans appear to be uniquely self-aware, and their minds are capable of projecting far into the future. For this reason, we humans must sit with death, our impending nonexistence, from the moment we are able to conceive of it. This understanding can be a powerfully frightening experience, as Ernest Becker, an American anthropologist whose work formed the theoretical foundation for TMT research, describes:

> Man is literally split in two: he has an awareness of his own splendid uniqueness in that he sticks out of nature with a towering majesty, and yet he goes back into the ground a few feet in order blindly and dumbly to rot and disappear forever. It is a terrifying dilemma to be in and to have to live with.[13]

Thus TMT researchers study the impulses and cognitions of subjects pressed with the fear of death, which they experimentally prime by a procedure called *mortality salience induction*. Subjects either read or write essays about death, usually their own death—in contrast to controls, who may be instructed to read or write relatively innocuous essays about things like television or food.

TMT research finds that when fears of death are primed, subjects more closely huddle around their political leaders, as we would expect from animals that evolved in hierarchical societies in which dominant individuals (typically males) functioned as protectors. One study conducted during the presidency of George W. Bush found that subjects significantly increased their support for the president and his antiterrorism policies when mortality salience was induced, relative to controls (who were asked questions about watching TV). A separate experiment in the same study found that reminders of the 9/11 terrorist attacks on the World Trade Center had the same impact as mortality salience induction on support for Bush, irrespective of the subjects' political orientation.[14] The results suggest that the tendency to gravitate toward powerful leaders is independent of political affiliation, and thus potentially very ancient.

Indeed, seeking assurance from a dominant male protector is something Bush himself engaged in. When asked whether he sought his father's advice about the war in Iraq, he replied, "You know, he is the wrong father to appeal to in terms of strength. There is a higher father that I appeal to."[15] Becker summarizes this psychological pull:

> It is [fear] that makes people so willing to follow brash, strong-looking demagogues with tight jaws and loud voices: those who focus their measured words and their sharpened eyes in the intensity of hate, and so seem most capable of cleansing the world of the vague, the weak, the uncertain, the evil.[16]

This is not to make the argument that rallying behind male leaders is an evolutionary inevitability. It was a woman leader in the Philippines, Maria Corazon Cojuangco-Aquino, who led the People Power Revolution and toppled Ferdinand Marcos, which could not have happened without popular support. The country also suffered several natural disasters while under her tenure, including an earthquake, a volcanic eruption, and a typhoon, and the people rallied behind her. Sri Lanka also had a woman president during ten years of its civil war. Liberia elected Ellen Johnson Sirleaf in the wake of the Second Liberian War. Similarly, Nicaragua rallied behind President Violeta Chamorro in the wake of a devastating civil war.

Likely, the circumstances in which women rise to power also reflect our evolutionary psychology. While primate males have protected their offspring from dangerous animals, females have protected their offspring from dangerous males, and today such roles can also be intuitively powerful in politics and religion. Chamorro, for instance, was seen as a powerful, protective mother figure and a martyr who stepped in and unseated Daniel Ortega, a man who critics saw as power-hungry despot.[17]

Yet the relative rarity of these protector-female historical figures points to an ancient history of male protectors, played out on an African savanna where protection required size, strength, and aggressiveness. Research finds that people prefer taller leaders and those with masculine facial features—which are produced by testosterone, a hormone associated with aggression, size, and strength—and that this preference is pronounced in the context of war.[18] Research has also found that the appeal of masculine leaders extends across cultures and that humans prefer masculine features in female leaders as well.[19] However, because we see masculine traits most commonly and most pronounced in men, the prevalence of protector-male political figures may reflect both our lingering primeval concerns about outside dangers and our unconscious preference for those traits most likely to succeed in the fight for our safety. The same may be true for the prevalence of protector-male religious figures.

Understandably, TMT research finds that fear of mortality influences subjects to more strongly endorse belief in God (and report greater religiosity). In one study the experimenters asked subjects to write an essay about what they believe happens to them when they die (controls were asked to write about their favorite foods). When death fears were primed, subjects endorsed greater religiosity and believed more strongly in God.[20] Numerous studies have found that mortality salience induction causes an increase in belief in God and the afterlife.[21]

While humans may have inherited their attraction to powerful male gods in times of danger from their primate ancestry, they appear unique in their ability to represent their evolutionary imperatives in the form of abstract concepts. The notion of an afterlife is one such concept and fits with TMT's supposition that humans build defenses against existential terror. What better way of assuaging fears of death than with the promise of eternal life? And as we might expect, it is a dominant male savior who provides not only protection from death by predation but protection from death itself. This is a fundamental dogma of Christianity, and references to a masculine savior and the afterlife are continual throughout the Bible:

> When the perishable has been clothed with the imperishable, and the mortal with immortality, then the saying that is written will come true: "Death has been swallowed up in victory." "Where, O death, is your victory? Where, O death, is your sting?" (1 Cor.15:54–55)

> Brothers, we do not want you to be ignorant about those who fall asleep, or to grieve like the rest of men, who have no hope. For we believe that Jesus died and rose again and so we believe that God will bring with Jesus those who have fallen asleep in him. (1 Thess. 4:13–14)

> After that, we who are still alive and are left will be caught up together with them in the clouds to meet the Lord in the air. And so we will be with the Lord forever. (1 Thess. 4:17)

Jesus said to her, "I am the resurrection and the life. He who believes in me will live, even though he dies; and whoever lives and believes in me will never die. Do you believe this?" (John 11:25–26)

> He will swallow up death forever. The Sovereign Lord will wipe away the tears from all faces; he will remove the disgrace of his people from all the earth. The Lord has spoken. (Isa. 25:8)

There is even reference to Christ protecting against predation (from the devil) and in the process giving everlasting life and assuaging mortality fears all at once:

> Since the children have flesh and blood, he too shared in their humanity so that by his death he might destroy him who holds the power of death—that is, the devil—and free those who all their lives were held in slavery by their fear of death. (Heb. 2:14–15)

Notions of the afterlife are seen across religions and cultures. The ancient Egyptians had the fields of Aaru, and the ancient Greeks and Romans had the underworld. The Norse cultures had Valhalla, Hel, or Nifhel. The Hopi tribe has the land of the dead. Buddhists, Hindus, and Wiccans have reincarnation. Muslims have paradise. Catholics have heaven, hell, or purgatory. The Mormons even stake claim to the planets of distant galaxies. It follows that powerful males are overrepresented as rulers of these afterlife territories: Aaru is ruled by Osiris, the underworld by Hades, the land of the dead by Masau'u, Valhalla by Odin, heaven by God, paradise by Allah, and hell by Satan.

PATERNAL CERTAINTY IN APES, MEN, AND GOD

The rules of natural selection come into play in the business of protection, which we can trace from apes to men to god. Because organisms are programmed by selfish genes to engage in kin altruism, a good means of assuring the support of a powerful individual is to be that individual's offspring. For males, giving support often means providing protection from danger. This can easily be observed in nonhuman primates. Wild savanna baboons, for example, will selectively support their offspring in agonistic encounters, which serves as protection from stress and injury and aids in rank acquisition.[22] Other male primates will selectively protect their offspring from infanticidal males.[23] Conversely, not being the offspring of a powerful male can increase the risk of infanticide; for fitness reasons, discussed more thoroughly in the next chapter, many male primates (including humans) will kill the offspring of rival males. In short, the many dangers of the natural world can make paternal affiliation a matter of life and death.

Given an evolutionary history in which affiliation with males brought protection from danger, and in which males preferentially support their offspring, it is understandable that people would create dominant male gods and seek to establish filiation with them. This tendency is strong in the Judeo-Christian creed, and many Bible passages assert God's paternity: "Ye are the children of the Lord your God" (Deut. 14:1); "All of you are children of the most High" (Ps. 82:6); "Ye are the sons of the living God" (Hosea 1:10); "We are the offspring of God" (Acts 17:29). Further, this relationship is tinted with human emotion; just as primate fathers experience love for their offspring, God is said to feel paternal love, notably in recognition of his followers' filiation: "See what great love the Father has lavished on us, that we should be called children of God!" (1 John 3:1).

Like male primates, God—the Father—intervenes in conflicts between his offspring and their competitors, which impacts his offspring's rank status: "For those who are led by the Spirit of God are the children of God. The Spirit you received does not make you slaves, so that you live in fear again" (Rom. 8:14–15). God supporting his offspring in conflict, particularly in battle (a topic I expand upon in later chapters), is a frequently recurring theme. For example, when God's descendant Joshua sets out to conquer Jericho, God confirms his support, "Do not be afraid; do not be discouraged, for the Lord your God will be with you wherever you go." (Josh. 1:9). He provides similar support to the "sons of Israel" (here the Reubenite, the Gadite and the Manasseh tribes), "They cried out to God during the battle, and he answered their prayer because they trusted in him. So the Hagrites and all their allies were defeated" (1 Chron. 5:20). In Christianity, the father-child relationship is concerned with care and protection; as Pope Benedict XVI explains, "The fatherhood of God, then, is infinite love, a tenderness that leans over us, weak children, in need of everything. . . . It is our smallness, our weak human nature, our frailty that becomes an appeal to the mercy of the Lord so that He manifest the greatness and tenderness of a Father helping us, forgives us [*sic*] and saving us."[24]

Paternal love may have its benefits for primates and primate-based gods, but it requires paternal certainty. A primate mother will always know that the child she gives birth to is hers. However, because ovulation and insemination are concealed in human females (and in many other primates), a male cannot always be sure that a child is his. For males, paternal certainty has implications for evolutionary fitness, for if a male is cuckolded, he risks spending a great deal of time and energy supporting another male's offspring. One means of establishing paternal certainty is by physical resemblance, and research shows that even nonhuman primates use visual cues to recognize kin.[25]

Resemblance ends up having important fitness implications in humans. For instance, research has found that men, more than women, are willing to invest more in hypothetical children (i.e., in pictures) whose faces are digitally composited with their own faces.[26] One study found that men (again, more than women) viewed children with faces that were digitally remixed with their own to be the most attractive; these were also the children they would be most likely to adopt, spend the most time with, spend money on, and have the least resentment about paying child support for.[27]

Facial resemblance also predicts *actual* fitness. A study of families in rural Senegal found that children's facial resemblance to the father predicted paternal investment and that this resemblance was related to better growth and nutritional status for the children, suggesting differences in resource provision.[28] The same study found that a child's odor similarity to the father also predicted the father's investment in that child, a behavior also seen in nonhuman primates who detect their kin by smell.[29]

Given that in our evolutionary history males selectively protected and supported their offspring and that paternal resemblance remains an important predictor of support among modern-day men, it should come as no surprise to find efforts to establish physical resemblance to God in scripture, based as he is on the dominate-male archetypes of our species. In Judeo-Christian creed, this effort is captured best in the doctrine of *imago Dei*, which holds that God created humans specifically in his own image. The book of Genesis establishes this resemblance: "When God created mankind, He made them in the likeness of God" (Gen. 5:1). The theologian Saint Augustine (354–430 BCE) of Hippo (present-day Algeria) was a highly influential religious thinker who has made his mark on Western thought, the stuff of contemporary college courses in philosophy. He offers a corporeal take on *imago Dei* (here, as cited by Saint Thomas Aquinas):

> [T]he body of man alone among terrestrial animals is not inclined prone to the ground, but is adapted to look upward to heaven, for this reason we may rightly say that it is made to God's image and likeness, rather than the bodies of other animals.[30]

The paternal resemblance of Jesus to God is also emphasized in the Bible, lest his filiation to God be uncertain (italics mine):

> The Son is the image of the invisible God, the firstborn over all creation. (Col. 1:15)

> The Son is the radiance of God's glory and the *exact representation* of his being, sustaining all things by his powerful word. After he had provided purification for sins, he sat down at the right hand of the Majesty in heaven. (Heb. 1:3)

The concept of *imago Dei*, a central dogma of the Judeo-Christian faiths, places humans in a different category from all other life, elevated as the intended product of God's creation itself. *Imago Dei* begets the doctrine of *man's dominion*—the idea that man, largely because he was created in the image of God, possesses the divinely conferred privilege of dominating all other life on earth (including women):

> Then God said, "Let us make mankind in our image, in our likeness, so that they may rule over the fish in the sea and the birds in the sky, over the livestock and all the wild animals, and over all the creatures that move along the ground." (Gen. 1:26–28)

> Be fruitful and multiply; fill the earth and subdue it; have dominion over the fish of the sea, over the birds of the heavens and over every living thing that moves on the earth. (Gen. 1:28)

God, in other words, has left men in charge. Saint Augustine explains why, linking our godliness to the soul:

> Man's excellence consists in the fact that God made him to His own image by giving him an intellectual soul, which raises him above the beasts of the field.[31]

As a result of *imago Dei*, God raises mankind above all other animals in the world as his special, protected progeny. This arrangement brings particular comfort and satisfaction to animals that evolved in dangerous environments, fearing the monsters crouched in the shadows. In the following quote, God places humans at the top of the food chain and reverses the order of fear and dread. Now, with his help, all animals of the world fear humans:

> The fear and dread of you will fall on all the beasts of the earth, and on all the birds in the sky, on every creature that moves along the ground, and on all the fish in the sea; they are given into your hands. (Gen. 9:2)

God's favoritism is based on primate sociality; dominant primates favor their offspring, and favoritism helps in rank acquisition. In nonhuman primates, inherited maternal rank (often with the help of mothers intervening in conflicts) is far better documented than inherited paternal rank,[32] while in humans, sons gaining the status and power of their fathers is more pronounced. Two recent examples of this power and status transfer in world politics—George W. Bush from George H. W. Bush, and Kim Jong Un from Kim Jong Il—add to a long history of sons (over daughters) inheriting the thrones of kings. Therefore it seems logical that if we see ourselves closely related to the most powerful being in the universe, we should inherit some power by filiation. If God were not modeled after a genetically programmed human, then we would not expect him to show favoritism based on selfish genes—but *man's dominion* claims immense power bestowed upon man by his father God, effectively imputing kin altruism onto God.

While being God's offspring has its benefits, particularly for sons, not being recognized can be a dangerous affair, much as it is in primate societies. In the wild, dominant male primates have been observed to systematically kill offspring that are not theirs.[33] Primatologists have been able to experimentally induce this behavior by removing the alpha male, at which time all infants are predictably killed by unrelated males, often those assuming the dominance role.[34] Similarly, the lack of facial resemblance is related to infanticide in humans; men who have committed infanticide often cite the lack of resemblance of the murdered child.[35] There are parallels for this, too, in the Christian tradition.

Christian theologians have interpreted the fall from grace—when Adam and Even disobeyed God's interdict against eating from the tree of knowledge—as the point at which mankind lost *imago Dei*[36] (which for Christians is restored by Christ). After Adam and Eve were no longer recognized as God's likeness, God caused them both immense pain and in effect killed them both by making them mortal, actions we might expect of a dominant male. Thus humans have created religious dogma that strives to establish paternal resemblance to God, fully consistent with an evolutionary history in which powerful males killed offspring not recognized as their own.

However, not all useful associations with dominant male primates are between father and offspring. Nonfilial relations between adult males have important fitness implications as well. Chimpanzees, for example, establish affiliation through grooming. Research shows that not only do adult males trade grooming for support in conflict[37] but that grooming is often directed up the hierarchical chain.[38] Higher-ranking males are generally more powerful and better able to assist in conflict; therefore, selectively seeking their alliance can produce greater fitness benefits for lower-ranking members.

Association with higher-ranking males also aids fitness in other ways. For example, dominant male chimpanzees will form coalitions, share mating, and block other males from their shared sexual claims.[39] Accordingly, reproductive success is often correlated to rank status, or *hierarchical proximity to the dominant male.*[40] Powerful men engage in similar behaviors with their number of wives often correlating to their proximity to the headman or king atop the social hierarachy.[41]

PROBLEMS OF DIVINE ALLIANCE-MAKING

In later chapters I will elaborate on the fitness advantages gained by men who ally themselves with a dominant male god. Here, however, I wish to explain the problems inherent to this particular affiliation. First, in the eyes of science, the existence of God is an untestable hypothesis yet (and very unlikely ever) to be empirically demonstrated. Likewise, in the words of religious dogma, God is untouchable, existing as he does in a different dimension. Therefore, all the tangible fitness advantages of making male alliances with God are exclusively enjoyed by mortal men. However, men—being male primates—will often use their purported association with God to serve their own evolutionary interests, often through tyranny and despotism, causing immense human suffering in the process. Doctrines such as *imago Dei* serve to give divine legitimacy to the autocratic decrees of powerful men. And because of our evolutionary fear of dominant males, such legitimacy is often protected from scrutiny, allowing men use to their acquired immunity to speak directly for God (or directly *as* a god).

Men have historically clamored for this position, motivated by its immense rewards of power. Walter Burkert, a German scholar of Greek mythology, gives a nicely detailed list of powerful men throughout European history who attributed divine authorization to their reign: Greek kings proclaimed that their authority came from Zeus; Alexander the Great claimed to have been the son of a god—Zeus Ammon; in Christian Rome, rulers were installed by *Dei gratia*, or "by the grace of God"; and the golden mosaic at Palermo depicts Christ placing the crown on King Roger's head. As Burkert points out, "The gods stand behind those who exercise worldly power; conversely the monarch is 'the head, immersed in prayer.'"[42] Because there is no tangible being with which to share power, there are some traditions in which men (and women) have simply become gods themselves. For example, the pharaohs of ancient Egypt were considered walking gods on earth, as were the rulers of the ancient Mayan, Aztec, and Incan dynasties.

The pope is another example of a veritable god on earth with the power—across long stretches of history—to force kings to grovel at his feet or to simply remove kings from power. He is considered the Vicar of Christ, or *the earthly representation of Christ and God*. This claim deserves further reflection, particularly given the sheer scope of power embodied by stations such as the papacy. Among religions, the Catholic Church stands out among the biggest economic and military juggernauts in religious history. The pope is at the helm of this staggering powerhouse. And popes are men who not only serve as deputies of the most powerful dominant male in the universe but who claim to speak directly for God.

One might argue that the enthusiasm with which men have assumed this position has been religiously motivated. However, a peek behind the gold-threaded papal robes reveals the hirsute forms of our primate male ancestors competing for positions of power in the ways of old, enflamed by violence and sexuality. One example comes from the tenth century, when Pope John XII was reputed to have turned the papal palace into a personal brothel. He was a notorious womanizer and was charged by King Otto I of Germany with committing adultery with his father's concubine along with several other sexual indiscretions. John was also charged with castrating his cardinal subdeacon—a behavior, as we will discuss, that apes and powerful men perform to take their competitors out of mate competition. King Otto came to Rome, deposed John, and appointed Pope Leo VIII in his stead; but when Otto returned to Germany, John slaughtered the leaders of the Imperial Party in Rome and repositioned himself in the papacy.[43] Falling victim to the same matecompetition ethos that compelled him to castrate and murder men and to monopolize women (behaviors we would expect from a dominant male primate), John was reportedly killed by a jealous husband while having sex with the man's wife.[44]

Also in the tenth century, Roman nobleman Bonifacio Francone strangled Pope Benedict VI and made himself pope. Soon after, he absconded to Constantinople, his pockets stuffed with the papal treasury. But Francone's alpha-ambitions didn't stop there; after a time, he returned to kill Pope John XIV and assumed the papal throne again, which he held until he died in 985.[45]

Earlier in the tenth century, Pope Sergius III ordered the murders of his predecessors Pope Christopher and Pope Leo V, which cleared the way to his papacy. He was also known to have a host of mistresses. Moreover, as God's favored son, Sergius' own (illegitimate) son eventually become Pope John XI.[46] Pope Alexander VI (reining from 1492-1503) was known for his fondness for orgies and had eight children from three or four mistresses. Although most of his sexual indiscretions took place before he assumed the papacy at the late age of 61, once there, he was keen on waging war on his political rivals and amassing a profane fortune in the process.[47] The list goes on, and although the similarities are somewhat obscured by religious dogma and perhaps the shimmering of papal jewels, one could just as easily place these machinations in the rainforests of Gombe.

Backed as it were by God, this astounding power of the papal office (variously used to take women, wealth, and the lives of male rivals), was no doubt inspired by Christ's own vicarship—that is, Christ's claim to be the sole representative of God. The inestimable rank of Christ is well established in the Bible:

> How can you say, "Show us the Father"? Don't you believe that I am in the Father, and that the Father is in me? The words I say to you I do not speak on my own authority. Rather, it is the Father, living in me, who is doing his work. Believe me when I say that I am in the Father and the Father is in me. (John 14:9–11)

Not unlike the pharaohs, Christ goes further and conveys his direct connection to God by saying, "I and the Father are one" (John 10:30). In addition, Christ makes it clear that his followers will not jump the chain of command, "No one comes to the Father except through me" (John 14:6). He also asserts that those who fail to love him, to recognize him as the embodiment of God, and to agree with his words, relinquish their filiation with God to become sons of Satan:

> "We are not illegitimate children," [the Jews] protested. The only Father we have is God himself." If God were your Father, you would love me, for I have come here from God. I have not come on my own; God sent me. Why is my language not clear to you? Because you are unable to hear what I say. You belong to your father, the devil, and you want to carry out your father's desires. He was a murderer from the beginning, not holding to the truth, for there is no truth in him. . . . Whoever belongs to God hears what God says. The reason you do not hear is that you do not belong to God. (John 8:42–47)

Of note, however, popes and other men who share alliances with man-based gods have not used such associations solely for small-scale, personal fitness gains as seen in the examples above; they have also martialed armies to slaughter their way across human history. Napoleon Bonaparte once said that "to honor the Emperor is to honor God Himself. . . . if they should fail in their duties to the Emperor . . . they would be resisting the order established by God . . . and would make themselves deserving of eternal damnation."[48] It is estimated that five to seven million people (mostly civilians) were killed during the Napoleonic wars.[49]

From the tradition of god-kings (and the occasional god-queen) to the modern-day American evangelicals who reign over megachurches, affiliative power is enduring. Evangelical John Hagee, for example, held weekly meetings with George W. Bush during his presidency. Bush in turn commanded the most powerful military force in human history, guided by his religious beliefs. As president, he led that force into war, what he called a "crusade against terror," reigniting a centuries-old religious conflict between Islam and Christianity. There have been numerous reports that in private meetings with statesmen or religious leaders, Bush claimed that his decisions as commander in chief were directed by God and that God spoke through him.[50] Furthering Bush's purported alliance with God, Lieutenant General William Boykin, a man central to the war against terror, asked while addressing Christian groups, "Why is this man in the White House? The majority of Americans did not vote for him. He's in the White House because God put him there for a time such as this."[51]

However, critical thinking must be employed in situations where powerful men claim to be speaking for God or acting as God's proxies on earth. We must understand that while claiming to be following divine mandates, men actually follow biological ones, which influence them to kill rivals in warfare. Further, men who claim divine inspiration for commanding armies to war also pronounce their roles as protectors, thus playing on evolved mechanisms geared to elicit ancient and powerful emotions in their subordinates. Such roles are also conflated with God's. For instance, General Boykin also proclaimed that the war on terror led by Bush was in fact a war on Satan, the animalistic predator embodying those "ghosts" of our ancestors' pasts. Similarly, Muslim extremists who commit acts of terror against the United States often portray America as the "Great Satan."

The power of evolutionary knowledge is that it unveils the designs of dominant men and allows us to rend apart those conflations with God that create enmity and bloodshed. It allows us to see that divine justification for destructive human behaviors by necessity exploits ancient fears, compelling us to huddle around men and man-based gods, often while in the process of starting wars. Perhaps with this knowledge we might make more rational political decisions, including whether we rally in support of political leaders in conflict, thus differentiating real, contemporary dangers from those feared by our ancestors.

Chapter 4

SEXUAL DOMINANCE: FROM APES TO MEN TO GODS

God appears to want our women. Despite his resplendent powers, he seems to find himself cloistering women away, monitoring their sex lives, and enacting violent revenge against their sexual independence. We also see him punishing the sexual ambitions of lesser males. Such ordinary sexual concerns do not easily reconcile with the notion of an everlasting being, one that requires neither women nor sex to reproduce itself into future generations. At his worst, we have a dominant male god behaving like a lustful, jealous, sexually acquisitive male primate. The clearest explanations for divine oxymora such as these lay within our own reproductive psychology. Understanding this point, and exposing Gods purported sexual concerns for what they really are, does much to shed light on religious sexual repression and religiously motivated violence against women.

APES

Violence and Sexual Access

Females are often in high demand, and this demand is tied to our reproductive biology. The biological limit for reproduction in female mammals is defined by gestation and lactation, whereas the limit for males is defined by the number of receptive females to which they can gain sexual access.[1] Practically speaking, this means in a given population of a relatively even sex ratio, we will find a higher proportion of sexually receptive males than females. This, combined with the fact that females are choosier in selecting their mates, means that most male mammals engage in heated competition for access to females.

Violence is a prevailing strategy for winning access to females among primates, including the great apes. For instance, dominant male gorillas often use the threat of aggression to exert totalitarian rule. Notably in gorilla troops, a dominant male's supremacy confers to him exclusive sexual privileges with the group's females. Further, because total domination is common and rarely contested, gorilla societies tend to be relatively stable (i.e., plagued less by violence).[2]

By contrast, rank among the common or *robust* chimpanzee is more fluid, characterized by shifting political alliances and fluctuating alpha status, both of which are principally brought about by violence. Consequently, access to sex varies according to dominance status. This is not to ignore the sexual agency of female primates, including chimpanzees, which may copulate in public with a more dominant male, and then sneak off for rendezvous with lesser males. However, the ability of dominant male chimpanzees to monopolize sex is clear, as is the use of aggression to establish and maintain the monopoly.

In his study of the chimp colony at Burgers' Zoo in Arnhem, Netherlands, Dutch primatologist Frans de Waal observed a particularly dominant alpha male named Yeroen. Yeroen's rule was absolute. At one point while in his position of power, he managed to secure nearly 100 percent of the mating in the whole colony. When Yeroen was finally deposed, none of the other chimps were able to achieve such total domination over their male troop-mates. Among remaining rivals, mating was highly correlated to the amount of dominance a particular male was able to impose at a given time, which varied according to alliances and fights won or lost. Observing this led de Waal to draw an important conclusion:

> There is, generally speaking, a definite link between the rank of a male and his copulation frequency, although it is by no means a rigid law but rather a rule to which exceptions are possible. It is not that high-ranking males are more virile, but that they are incredibly intolerant and chase lower-ranking rivals away from estrus females. If they catch another male mating, they intervene by attacking him or his mate. Females are also clearly aware of this risk.[3]

Male chimpanzees, then, may be forced to demonstrate submission to the dominant male by ceding reproductive access to him. Females may be forced to submit only to the dominant male or face violent reprisal. The more powerful the male chimpanzee, the more he is able to sexually monopolize females.

This use of violence as a reproductive strategy is reliably seen among other primates as well. In his book *Macachiavellian Intelligence*,[4] primatologist Dario Maestripieri gives an insightful account of macaque mating behaviors. These monkeys live in matrilineal societies in which females hold more power. While male macaques emigrate from their natal groups, females typically remain, allowing them to amass alliances and power in their home territory over time. Nevertheless, even in these matrilineal societies, dominant males manage to enact the usual strategies—winning violent competitions for sexual privileges with numerous females.

High-status females will sometimes mate with several males. They may mate with lesser males when less fertile, which may be an attempt to confuse paternity or to secure favors such as grooming. Even so, knowing the alpha's reputation for violence, the deed is almost always done quickly and on the sly, away from the eyes of the dominant male. But when high-status females are ovulating, they show a strong preference for mating with dominant males.

If there is only one female in estrus, the dominant male macaque may be able to monopolize her completely, but oftentimes many females come into estrus simultaneously. Surrounded by "estrus females," the alpha male may become something of a tyrannical control freak, pressed with the constant, exhausting task of patrolling his numerous sexual claims and chasing off and punishing his sexual rivals. This can become a veritable marathon of sex and violence. Notably, female transgressors are punished as well. If the dominant male catches a female mating with another male, he usually attacks her. In fact, like brutish men, macaques are known for attacking females for mere flirtations, such as grooming other males. Macaque females use symbolic sexual submission to avoid attacks, briefly presenting their hind quarters to the dominant male. To avert the alpha's wrath, lesser males may also use this gesture.[5]

Many baboons follow similar patterns of sex, violence, and submission. Hamadryas, gelada, and Guinea baboons live in groupings of females—aptly called *harems*—that are dominated by a single male that enjoys (virtually exclusive) sexual privileges with the entire group.[6] Male gelada and hamadryas baboons win dominance through violent competition with other males and are known for staging raids on other males' harems to steal females.[7] Sometimes coalitions will form between lesser males seeking to overthrow the dominant male for access to females. In order to protect their sexual monopolies, dominant hamadryas males are known for herding females by making threat displays and chasing, restraining, dragging, or biting the necks of those who stray too far or approach lesser males.[8] Like macaques, lower-ranking male baboons may present their hind quarters to appease the dominant male, and in response, the dominant male may also briefly mount the submissive male[9]—a behavior sometimes called a *false mount*.

The relationship between dominance and sexual access is not unique to chimps, gorillas, baboons, and macaques; while differences in mating patterns exist, most primates engage in some variation of the machinations noted above. Even in bonobos, a subspecies of chimpanzee often characterized as "egalitarian" and "peaceful," dominance rank exists among males,[10] and copulation frequency has been found to correlate to dominance.[11]

Finally, in skirmishes, nonhuman primates spend a lot of time attacking each other's genitalia. This has been observed in macaques, baboons, and chimpanzees in both the wild and in captivity,[12] and such assaults often result in castration. The genitalia are indeed a vulnerable place to attack, with the benefit of inflicting a great deal of pain. Literally castrating another male has the added advantage of removing him as sexual competition altogether.

Infanticide in Nonhuman Primates

Infanticide is an especially dark male strategy, one that has been observed across countless mammals, including numerous monkey species, all of the great apes, and humans. This behavior is almost never performed by females. Infanticide typically occurs when one male gains supremacy by overthrowing another. However disturbing, the reproductive advantages of infanticide are clear. For one, by killing the offspring of rival males, the dominant male eliminates future competition that may impinge on his fitness or that of his offspring. Accordingly, nonhuman male primates will selectively kill the offspring of other males while protecting their own from infanticide.[13] Second, in many primate species, a mother immediately goes into estrus when her infant is killed. In an uncertain world of tenuous survival, time is critical and raising dependent infants, weaning them, and returning to fertility is time-consuming. Males who kill rival offspring and quickly reproduce with their mothers have an advantage in the race for survival.

American anthropologist Sarah Blaffer Hrdy points out that this behavior can become something of an evolutionary "trap" for both sexes.[14] Though counterintuitive, female animals stand to gain from mating with infanticidal males because those males tend to produce infanticidal sons. Having male offspring that are not infanticidal in a world of infanticidal males would be an evolutionary dead end—the genes of those offspring would be wiped out of the gene pool by infanticide. Nevertheless, while the female may make evolutionary gains by mating with the infanticidal male, the death of her offspring, whom she has undergone the risk to bear and nurture, is an unequivocal loss. The obvious stress and anguish involved is another cost.

Primatologists have frequently made observations of infanticide in the field. Below is one account in chimpanzees. Here, two males attacked a female and her infant. The female fled, and one of the males grasped the distressed infant.

> Its nose was bleeding as if from a blow and [the male], holding the infants legs, intermittently beat its head against a branch. After three minutes, he began to eat the flesh from the thighs of the infant which stopped struggling and calling.[15]

The primate behavior patterns we have been discussing clearly emerge in the dominance struggles of men—for example: sexual monopolization, male and female sexual submission, men castrating other men, and even infanticide. As among our primate cousins, the more powerful the man, the more he is able to successfully enact these strategies.

MEN

What Men Want

In slow motion, an extremely sexy, young brunette woman runs barefoot through the forest wearing nothing but a skimpy brown bikini. Her hair is messy. She looks primitive, animal-like. Instantly your mind goes to the sexual implications. She turns to see other brown-haired women, also bikinied, young, and beautiful, all running in the same direction. The scene quickly cuts to a young blonde woman, running up a hill. The blonde is also gorgeous and fit, and you see that she, too, is followed by other blonde, bikinied women running up the hill. The scene pans to aerial view, revealing droves of young blonde women streaming through the valleys and hills, not unlike an aerial view of a battle scene. Just then we see waves of black-haired women diving into the sea and riding toward land on enormous tsunamis. There is a crescendo. We can see that the gorgeous women of the world are coming from every direction to where they will meet on a beach. The camera pans to a skinny young man dousing himself with some sort of body spray. The women seem to be attracted to the spray and start rushing more excitedly, although now it becomes clear that what they really lust for is the man. Luscious young women are charging the man from all sides, filled with desire. The sexual possibilities are endless.

This scene is from an Axe® body spray commercial, which follows the scene with the phrase "spray more, get more." While obviously hyperbolic, the commercial seems to resonate with men by targeting sexual fantasies driven by an evolutionary numbers strategy. Studies overwhelmingly bear out the male preference for quantity, which can be measured by the desire for casual sexual partners. Research on sexual fantasies has found that men strongly prefer casual sex more, masturbate more,[16] and fantasize about group sex more than women.[17] One study found that men have a higher daily frequency of sexual fantasies and, not far off from the Axe® spray commercial, were four times more likely than women to report sexual fantasies with more than one thousand different individuals over the course of their lives.[18] In addition, a set of studies had attractive undergraduates approach the opposite sex and quickly offer casual sex. The researchers found that a whopping

75 percent of men accepted the request, as compared with none of the women.[19]

Men and women also differ in the perceived importance of sexual infidelity. In addition to casual sexual partners, many men also want long-term relationships.[20] While this seems counterintuitive, it fits with the fact that males will invest in women and in the children they have together with them. This often brings male concerns over infidelity to a head. For example, men in one study were distressed by the thought of their partner having a sexual affair with a man (thus risking pregnancy), but were not distressed by the thought of their partner having a sexual affair with a woman.[21] Further, a number of studies across cultures have found that men display more distress at the thought of sexual infidelity, whereas women display more distress at the thought of emotional infidelity.[22] The general explanation for these findings is that men evolved fear of cuckoldry (which is not a risk for women), and women fear that emotional connection to other women will result in resource diversion. The end result is that, like other male primates, men guard their mates.

Based on evolutionary logic, it would seem that the ideal male fantasy would involve a wide variety of women whose sexual loyalty could be guaranteed. When resources (and laws or religious customs regarding polygamy) have permitted, history shows how eagerly men have competed for this ideal.

What Dominant Men Get

To a greater or lesser extent, every human society is organized by status hierarchy, although our evolved capacity for complex culture has clearly expanded since our proto-human forebears physically fought their way into positions of power. Today, reaching high rank among men no longer requires violence or intimidation to the extent it did in the past. Even so, violence and intimidation are still often used, though more commonly in subtler forms consistent with contemporary social mores. Regardless of whether it is achieved by dominance, or more peaceably by prestige, high rank continues to be characterized by control (or potential control) over resources—which women have traditionally been attracted to for the reasons discussed in chapter 2.

The research on male dominance, wealth, and reproductive success is clear. A cross-cultural study of 104 societies, spanning every continent on the globe, found that dominant men overwhelmingly possessed greater wealth and more mates.[23] In hunter-gatherer societies, status is often gained by direct aggression and raw physicality. Yąnomamö tribesmen of the Amazon, for example, gain status through chest-punching duels, ax fights, and by killing rivals from other groups in combat. Men in this society who have killed other men in battle have more wives and more children than men who have not.[24] With no currency or means of preserving food, accumulated wealth differs little between men in Yąnomamö society. However, dominant men often receive food tributes from other families in their village. This increases their overall access to material resources, which they use to support more women and children.[25]

In modern societies, the means of acquiring resources have moved from raids and hunting to more complex forms of acquisition, but the relationship between rank and sexual success remains robust. The obvious winners are the rock stars, the billionaire playboys, the professional athletes, and the Hollywood actors. Gene Simmons, front man for the American rock band Kiss, once claimed to have had sex with 4,800 women, as one example.[26] American evolutionary psychologist David Buss's studies speak to this trend, showing that financially successful men typically gain more sexual access to women.[27]

Differential access to any resource typically creates competition, but competition for mates is driven by special urgency. Most critically, no mating equals no reproduction, and unsuccessful men have a dangerous lack of sex. Though metaphorical, calling this dangerous in evolutionary terms isn't far off the mark. Never reproducing can have the same effect as being killed—it is by and large an evolutionary dead end (with the exception of genes aided by kin altruism, as described in chapter 2). Evolution has accounted for this by a familiar means of dealing with danger—*aggression*.

In her book *Despotism and Differential Reproduction*,[28] anthropologist Laura Betzig offers a highly generalizable summary of the phenomenon the title so succinctly describes. Drawing data from a large cross-cultural sample developed by Murdock and White,[29] Betzig studied the relationship between despotism and differential reproduction in 104 human societies worldwide. She carefully operationalized despotism as "the exercised right of heads of societies to murder their subjects arbitrarily and with impunity."[30] Her hypothesis was also clear: "Hierarchical power should predict a biased outcome in conflict resolution, which in turn should predict size of the winner's harem, for men, a measure of success in reproduction."[31]

What Betzig found was an overwhelming tendency for male rulers exercising great power to utilize that power toward great reproductive ends by tyrannizing rival males into submission and monopolizing females—familiar patterns seen in nonhuman primates. Citing societies from every continent on the globe, Betzig shows that when men are in positions to behave despotically, they do so, and with great enthusiasm. In all cases, rulers created laws enabling them to act with impunity, while rule breakers among the populace were tortured and killed, or had their hands, limbs, or testicles removed. Numerous rulers legislated for the castration of male adulterers, thus permanently cutting off their rivals' ability to compete sexually. Methods such as these resulted in greater access to females—harems in Betzig's sample ranged from two to thousands of women.

Among the ancient Inca of Peru the near perfect correlation between rank and sexual privilege was methodically codified into law. Here Betzig references Poma de Ayala:

> Caciques or principal persons were given fifty women "for their service and multiplying people in the kingdom." Huno curaca (leaders of the vassal nations) were given thirty women; guamanin apo (heads of provinces of a hundred thousand) were allotted twenty women; waranga curaca (leaders of a thousand) got fifteen women; piscachuanga camachicoc (over ten) got five; pichicamachicac (over five) got three; and the poor Indian took whatever was left![32]

Census records among the Azande of Sudan indicate that for every one hundred adult men there were twenty-six bachelors, forty-seven men with one wife, eighteen men with two wives, and nine men with more than two. Access to women was a privilege correlated with male status. *Zande* chiefs (a different tribe), on the other hand, might have anywhere from 30 to 100 wives, and kings, over 500 women.[33]

In Dahomey (present day Republic of Benin in West Africa), the king also had numerous wives and concubines, numbering into the thousands. Here, sexual submission was codified to an interesting extreme. Should one of the king's concubines decide to go out among the villagers, her presence would be signaled by a bell-ringing servant. All the villagers were required to turn their heads away, and men were required to keep a further distance than were women. When the bell rang, people moved hastily into their submissive positions for fear of the consequences; Dahomey chiefs often dealt with the lustful offenses of lesser men by decapitating them, impaling them on posts, or killing and hanging them upside down to decompose near the marketplace—an unequivocal warning to would-be adulterers.[34]

Moreover, Betzig found that power not only confers quantities of women but virgins, too. Across history, virgins have been among the jewels gracing the harems of powerful men. Among the Inca, for example, only virgins were allowed into the king's harem. As we have discussed, the evolutionary strategies of men influence the value they place on female virginity. By acquiring virginal brides (concubines, etc.), men ensure that their women come untainted by the genes of rival males. In a world in which power has evolutionary implications, those with rank have the privilege of monopolizing virginal mates.

To further ensure fidelity, harems were fortified enclosures. However lush and serene on the inside, which they sometimes were, harems were typically walled fortresses hidden away in the very center of the palace with armed guards (most often eunuchs) strictly controlling movement in and out. It is hard to imagine a more direct means of controlling female sexual resources. Ganda kings (of Uganda) had their harems surrounded with high reed fences and guarded by trusted servants. Citing Anglican missionary and ethnographer John Roscoe,[35] Betzig notes that the guards "admitted passages on the pain of death, or of 'some terrible mutilation in the event of his life being spared.'"[36] One wonders how often the mutilation involved male genitalia.

GODS

The Lustful Godhead

The Greek gods were notoriously sexual. Following the tradition of primates, the most dominant of the Greek gods (Zeus) gained his power by overthrowing another dominant male (his father, Cronus), and in doing so won dominion over his territory (Olympus). From here, Zeus ruled over the earth and sky with a violent temperament, commanding fearsome thunderbolts as weapons, which he used to enforce his will on those beneath him. As with other dominant males, power and authority conferred sexual privileges; though Zeus was married to Hera, he fathered many gods with other goddesses and with Titans—for instance, the celebrated twins Apollo and Artemis were a product of his affair with Leto, the Titan, not Hera, his wife. Zeus was also known for having numerous affairs with *human* females: Io, Semele, Callisto, and Europa (whom Zeus abducted and raped), among others. The main point here is that Zeus was a womanizer like other powerful males. This is not an isolated occurrence, but rather a common pattern of conduct across the gods of many religions.

Krishna is considered a supreme being among gods in Hinduism. It is told that in early life he drew scores of young milkmaids into ecstatic "dances" in the forests. To the dances, the women brought offerings of food, jewelry, their clothing, and, naturally, their bodies. As the story goes, after a stretch of playing hard to get, Krishna had a six-month-long marathon of sex with all the milkmaids. The milkmaids fell into deep, ecstatic longing for Krishna, titillated as they were by his reputation as a powerful warrior. Notably, many of Krishna's sexual conquests were with married women, making their cow-herder husbands Krishna's cuckolds in the evolutionary competition for sex. Krishna aptly used his wiles and status as a god toward this end, and the women willingly gave up their lowly cow-herder husbands for Krishna, a more dominant male.[37]

In competition, Krishna did not rely on charm alone. When necessary, he slaughtered other dominant males to win control over their women and territory. For example, Krishna overthrew and killed his uncle King Kansa and took Kansa's subjects to establish his own kingdom. Krishna then married the princess, Rukmini, by abducting her from an arranged marriage with King Shishupala, whom he later killed. Knowing

Krishna's status, Rukmini enthusiastically welcomed this new arrangement. Krishna went on to kill Narakasura, a male demon, and in doing so won 16,100 maidens.[38] This was a spectacular prize, even by the standards of the most successful despotic kings. Like other gods, there is a great deal of complexity to Krishna's personality; he can show beneficence and kindness if he chooses, but his primate origins predispose him to engage in violence and sexual domination.

There are scores of patriarchs in the Old Testament who were polygamous, including: Abraham, Abijah, Ahab, Ashur, Belshazzar, David, Elkanah, Esau, Gideon, Hosea, Jacob, Jehoram, Joash, Jehoiachin, and Rehoboam.[39] Many of these men achieved their monopolies over women with help from their male alliance with God, for example: "I anointed thee king over Israel, and I delivered thee out of the hand of Saul; And I gave thee thy master's house, and thy master's wives into thy bosom" (2 Sam. 12:7–8). King Solomon, however, trumps them all, and rivals even the despotic men in Betzig's study, with seven hundred wives and three hundred concubines (1 Kings 11:1–3). Even the Abrahamic god himself has been known to womanize. In Ezekiel 23, Yahweh is described as having not one, but two wives Samaria and Jerusalem—actually entire cities, personified; a subject to which we will return.

There is much to be said about the similarities between Hinduism and the Greek, Roman, and Abrahamic religions. Evidence suggests that these similarities are not mere coincidence; rather they are a reflection of these cultures having had more historic contact with one another than previously understood, which has resulted in the dissemination of certain religious tenets between them. But there are numerous facets of religious tradition that are *not* shared between these cultures. Those that are shared with such vigor are those that touch upon our shared evolutionary psychology.

Sexually Repressive Gods: Divine Jealousy

Religion and sexual repression have historically gone hand in hand, and religion's sexually repressive dictates have trickled into the mores of cultures across the globe. To understand the origins of sexual repression we must consider where the phenomenon occurs in nature and whether there is an evolutionary motivation that is elaborated in human cultures. I argue that sexually repressive behaviors and their representative doctrines and ideologies are rooted in mental architecture designed for navigating the social world of our primeval ancestors.

In nonhuman primates, repression is invariably related to rank, occurring when those higher in rank strive to obstruct the sexual impulses of those lower. Dominant male humans enact similar strategies, such as when kings reign over harems and make eunuchs out of other men. And neither are gods immune to such primate desires and jealousies. In many religious contexts we see God's jealousy commanding that men and women stave off their sexual impulses toward others, directing their attention to him instead. During Ramadan and Lent, millions of Muslims and Catholics (respectively) abstain from sex as a sacrifice to God, as one widely manifested example.

For organic beings with finite life spans, jealousy provides important motivation to compete for survival resources, be they food, water, territory, or sexual partners. However, we must take a moment to recall that according to scripture the Abrahamic god is a being who survives eternally without resources as we know them—thus, having no need to eat, why fight over territory that provides a regular food source? Lacking the threat of death, why should there be such a great concern over sexuality? Further, he is described as all-powerful, as having created the entire universe and life on earth, as performing humanly impossible miracles through his will alone—God should be jealous of no one, for a truly omnipotent being should have no truly viable competitors. Nevertheless, the Abrahamic god is unequivocal about his demand for exclusivity:

> You shall not make for yourself an idol, or any likeness of what is in heaven above or on the earth beneath or in the water under the earth. You shall not worship them or serve them; for I, the LORD your God, am a jealous God. (Exod. 20:4–5)

Powerful men often demand exclusivity, in line with powerful nonhuman primates. Among these males, *sexual* exclusivity is of prime concern. Similarly, God's demand for unrivaled allegiance is often framed in terms of sexual jealousy:

> For thou shalt worship no other god: for the LORD, whose name is Jealous, is a jealous God: Lest thou make a covenant with the inhabit-

ants of the land, and they go a whoring after their gods, and do sacrifice unto their gods, and one call thee, and thou eat of his sacrifice And thou take of their daughters unto thy sons, and their daughters go a whoring after their gods, and make thy sons go a whoring after their gods. (Exod. 34:14–16)

Likewise, many contemporary religious thinkers posit that idolatry—worshipping symbolic representations of other gods—is the same as adultery. Others have argued that *spiritual adultery*—the notion that worshipping things over God, such as one's husband, one's wife, money, or other religions—is analogous to cheating sexually on God. Another passage from the Bible that speaks to the sexual aspect of commitment to God: "Now the body is not for fornication, but for the LORD; and the LORD for the body" (Cor. 6:13).

The Virgin and the King

Like other dominant males, the Abrahamic god shows a preference for virgins. The god of Christianity is purported to have had a relationship with Mary of Nazareth. Like Zeus, Krishna, and an endless succession of other dominant males, he took another man's mate. And like the conquests of Zeus, this god never asked Mary if she wanted the relationship; he simply took what he desired. Mary, of course, was said to have been a virgin. Now, as I've noted above, human primates prefer virgins because they reduce the threat of uncertain paternity. Yet an omniscient (or all-knowing) God should already know if a child is his, and an everlasting god, who does not need to sexually reproduce himself into future generations, shouldn't really care. Thus God should neither fear cuckoldry nor prefer virgins, unless, of course, he is based on a male human.

Perhaps most oddly, why is Mary said to have remained a virgin after God is described as begetting a son with her? This child develops in a literal, physical sense in Mary's womb and was birthed in the exact manner as other human babies, and most Christians concede that every step of this process abided by the biological rules of human reproduction . . . *except for the sex*. The sex part was somehow bypassed, omitted, or transcended. Interestingly, suggesting otherwise creates huge stirs of emotion, shouts of blasphemy, and feelings of perversion, disgust, and moral outrage. As concerns Mary, such responses have much to do with the evolutionary value placed on the chastity of women, which we have discussed. These reactions also demonstrate the puritanical background of modern Christianity (particularly in America) that would seek to flatly deny the sexuality of God.

The Mormons, on the other hand, have dispensed with the allegorical sex between Mary and God and opt for a more literal interpretation. In the Book of Mormon, God's value of virginity—and abhorrence of its converse, sexual autonomy—is expressed unequivocally: "For I the Lord God, delight in the chastity of women. And whoredoms are an abomination before me; thus saith the Lord of Hosts" (Jacob 2:28).

There are many other examples. The consecration of virgins is an enduring practice of the Catholic Church, where a woman is consecrated to a lifetime of virginity in the service of God. Surrendering female sexuality to the dominant male is actually codified by the Catholics. Below is an excerpt from the Church's Code of Canon Law:

> Similar to these forms of consecrated life is the order of virgins, who, committed to the holy plan of following Christ more closely, are consecrated to God by the diocesan bishop according to the approved liturgical rite, are betrothed mystically to Christ, the Son of God, and are dedicated to the service of the Church. (Canon 604)

Here we have virgins married off to Christ in the manner of kings. But was Christ a god-king in the manner of humans? Christ is faithfully known as the *King of Kings* and *King of the World*. And in keeping with human kings, the Bible has Christ requiring virgins, at least per Paul's description in Corinthians: "For I have espoused you to one husband that I may present you as a chaste virgin to Christ" (11:2). It is impossible to know what the historical Christ truly felt on this issue. There was so little written about Christ in his day, and what is written in the Gospels took place across centuries and changing political landscapes, dimming much hope for historical accuracy. Nevertheless, the persistent references to virginity, purity, chastity, and sexual restraint continue, perhaps teaching us more about the need for the legend of Christ than for the Christ of history.

The Koran is similar to the Bible in its emphasis on virginity and contains the most notable reference of this kind with the god of Islam promising virgins in paradise for his loyal followers. Christians also describe virgins in heaven, but in a crucially different context; rather than reserving virgins for God's followers, God's followers are reserved as *virgins for God.* Here Matthew describes Christ receiving his followers (of both sexes) like virgin brides: "Then shall the kingdom of heaven be likened unto ten virgins, which took their lamps, and went forth to meet the bridegroom" (Matt. 25:1). Matthew goes on to state that five of those ten virgins were not ready for Christ and were therefore not allowed into his territory. The passage is intended as a warning to those not immediately ready to submit themselves to Christ as virgins:

> The virgins who were ready went in with him to the wedding banquet. And the door was shut. Later the others also came. "Lord, Lord," they said, "open to us!" But he replied, "Verily I say unto you, I know you not." (Matt. 25:10–12)

The idea of Christ as a bridegroom and his followers as his bride is widely expressed in Christian belief, known aptly as the Bride of Christ. The brides of Christianity, whether wedded to Christ or to men, are often subject to the chauvinistic requirements of dominant men. Peter warns that God wants women to dress like virgins: "Whose adorning let it not be that outward adorning of plaiting the hair, and of wearing of gold, or of putting on of apparel" (1 Peter 3:3). The attire of men faces far less scrutiny of this kind. Virginity, chastity, purity, and the like are usually more-highly valued by men than by women, and dress is a means to outwardly advertise either sexual devotion or sexual availability.

Gods in religions worldwide share an interest in sex, and the male gods seek and acquire sex in patterned, dominant-male style, with a noted preference for virgins—females free of the genes of rival males. This emphasis on virginity extends well beyond the Abrahamic religions, which we would expect if a concern for paternal certainty originates from our common evolved psychology as humans. For example, in ancient Incan culture there were castes of young virgins dedicated solely to the sun god.[40] Like Incan kings, the sun god desired young women certain to *not* be carrying another male's offspring. East Indian religions share this preference. Like Jesus, Krishna was purportedly born of the virgin Devaki, who, because of her purity, was selected to become the mother of god—meaning that his father, Vishnu, also appreciated virginity.[41]

Chaste and Submissive

As we have seen in our study of apes, monkeys, and men, dominant males are often preoccupied with blocking the sexual ambitions of their subordinates. This behavior has its reproductive advantages. God also enforces sexual restraint in his subordinates, even though he should have no real competitors of any kind. This is seen across many religions where holy men are expected to be chaste, or to at least possess fewer wives than other men in their communities—with the exception of those instances when men blur the line between themselves and their dominant god, such as in Mormonism, where higher-ranking holy men have more wives and children.[42]

Accordingly, Catholic priests take vows of chastity. Like the vows of nuns, priest's vows are described by Canon law as enabling men to become closer to Jesus and God. They are, in essence, men vowing to remain sexually chaste in order to serve the prerogatives of a more powerful male, rather than to compete with him:

> Clerics are obliged to observe perfect and perpetual continence for the sake of the kingdom of heaven and therefore are bound to celibacy which is a special gift of God by which sacred ministers can adhere more easily to Christ with an undivided heart and are able to dedicate themselves more freely to the service of God and humanity. (Canon 277)

The Church's position has been unambiguous on this point. The Council of Trent, considered among the important ecumenical councils of the Catholic Church, met numerous times between 1545 and 1563 to promulgate church canon and define outlawed heresies. One aim was to excommunicate and curse priests (and nuns) who did not abide by the council's rulings:

> Whomever shall affirm that the conjugal state is to be preferred to a life of virginity or celibacy, and that it is not better and more conducive

to happiness to remain in virginity or celibacy than to be married, let them be accursed. (Canon 10)

In nonhuman primates, the alpha doesn't repress all sex, just sex with rivals. By default, this leaves sex with the alpha as the only option. Tellingly, religious men are prone to give themselves sexually to God, following the ancient primate legacy of male sexual submission. The Song of Solomon from the Old Testament is considered by theologians to be an allegory for the relationship of God to Israel, or for Christians to Jesus or God. The soul in Christianity has been described as female, thereby enabling both sexes to be brides of Christ and become sexually subservient to Jesus. The Song of Solomon is lengthy and goes on in salacious detail:

> Let him kiss me with the kisses of his mouth: for thy love is better than wine. (1:2)

> Because of the savour of thy good ointments thy name is an ointment poured forth, therefore do the virgins love thee. (1:3)

> I sleep, but my heart waketh: it is the voice of my beloved that knocketh saying, Open to me, my sister, my love, my dove, my undefiled, for my head is filled with dew, and my locks with drops of the night. (5:2)

> My beloved put in his hands by the hole of the door, and my bowels were moved for him. (5:4)

> I rose up open to my beloved; and my hands dropped with myrrh, and my fingers with sweet smelling myrrh, upon the handles of the lock. (5:5)

In nonhuman primates, submissive males often make sexual displays toward the alpha. It follows naturally to find men showing sexual submission to God. If male sexual submission to God is indeed metaphorical, as tends to be the stance of religious thinkers, then it is by definition a "false mount," as we see in other primates—it is meant to symbolize submission.

Another way to show submission to the alpha is to give his females wide berth. This is clear across primate species, and the rules are fairly stereotyped: stave off sexual impulses, get spared the alpha's wrath. Conversely: try to have sex with his females, *incur* the alpha's wrath.

In the book of Revelations, God rained down a horrid, violent, and spiteful fury upon the earth. First, God took all the fruits of the earth, put them into a giant winepress, and turned the pressings into blood. With them he flooded the earth and the cities with blood. The blood flowed high, "up to the bridles of horses." God then sent disease to all the people, producing festering sores all over their bodies, after which he killed every living thing in the sea by turning the sea into blood. For good measure, he then turned every river into blood as well. He then made the sun burn every person alive. Despite these tortures, the people did not show proper submission. In other words, they "did not repent and give Him glory," so God responded accordingly and turned the world dark and so full of pain that the people "gnawed their tongues because of the pain." But the people yet again "did not repent for their deeds." God obliged and thrashed them further; he dried up the Euphrates River and then sent great acts of nature to punish the people—hail storms, lightening, and earthquakes—that effectively leveled cities, mountains, and islands.

Why would God reap such drastic punishment? In the pattern of dominant primates, it was over a female. This violent tirade was punishment for his woman giving her sexuality to a rival male. Ancient Babylon was considered one of God's wives. When she engaged in fornication with a rival male, "the beast," God punished not only her but all those who would dare to join with his rival:

> And another angel followed, saying, "Babylon is fallen, is fallen, that great city, because she has made all nations drink of the wine of the wrath of her fornication." Then a third angel followed them, saying with a loud voice, "If anyone worships the beast and his image, and receives *his* mark on his forehead or on his hand, he himself shall also drink of the wine of the wrath of God, which is poured out full strength into the cup of His indignation. He shall be tormented with fire and brimstone in the presence of the holy angels and in the presence of the Lamb. And the smoke of their torment ascends forever and ever; and they have no rest day or night, who worship the beast and his image, and whoever receives the mark of his name." (Rev. 14:8–12).

Revelations goes on to explain the sexual nature of the indiscretion:

> Then one of the seven angels who had the seven bowls came and talked with me, saying to me, "Come, I will show you the judgment of the great harlot who sits on many waters, with whom the kings of the earth committed fornication, and the inhabitants of the earth were made drunk with the wine of her fornication." So he carried me away in the Spirit into the wilderness. And I saw a woman sitting on a scarlet beast which was full of names of blasphemy, having seven heads and ten horns. The woman was arrayed in purple and scarlet, and adorned with gold and precious stones and pearls, having in her hand a golden cup full of abominations and the filthiness of her fornication. (Rev. 17:1–4)

It is striking that, even among all this destruction and chaos, God elected to spare 144,000 men, on the basis of their good behavior. Given the moralistic focus of the Christian tradition, one might expect to see God rewarding traits such as industriousness or being helpful to those less fortunate, perhaps teaching prosocial behaviors to children, or honoring one's parents and family. But none of these things were ultimately as important as staving off sexual impulses and staying away from God's woman Babylon. He spares 144,000 males because they were *virgins*:

> These are they which were not defiled with women; for they are virgins. These are they which follow the Lamb whithersoever he goeth. These were redeemed from among men, being the first fruits onto God and to the Lamb. (Rev. 14:4)

This metaphor is an example of how alpha-male primates typically decide who in their society is allowed to mate and who is not—a society of freely mating subordinates would defy the rules of the hierarchy.

Finally, the most definitive means to surrender to God's sexual prerogatives is the same means to surrender to the despotic king's—through castration, whether literal or metaphorical. Matthew makes reference to this:

> For there are some eunuchs, which were so born from their mother's womb: and there are some eunuchs, which were made eunuchs of men: and there be eunuchs, which have made themselves eunuchs for

the kingdom of heaven's sake. He that is able to receive it, let him receive it. (Matt. 19:12)

Religious thinkers have again argued that castration as described here is really a metaphor for something like celibacy, or spiritual loyalty. However, the act of castration has proven so common among despotic men and apes that we are left letting Matthew's words speak for themselves.

WOMEN

What Women Want in Their Men and Gods

Women tend to be drawn toward powerful male gods, just as they are attracted to powerful men. For women, Henry Kissinger famously said, "power is the ultimate aphrodisiac."[43] The extant research literature strongly supports this observation. For instance, one study of 5,000 American college students found that women regarded "status, prestige, rank, position, power, standing, station and high place"[44] significantly more important in a potential mate than did men.

Today, this notion of power may be more related to prestige than dominance, but women's attraction to big men (as we learned in chapter 2) reveals that our "modern skulls house a stone age mind."[45] The research on this female preference is clear. During the most fertile phases of their menstrual cycles women tend to prefer taller men,[46] more masculine body types,[47] more masculine facial features,[48] and men displaying competitive behaviors (e.g., derogating potential rivals).[49] During fertile periods, women also rate the odors of men scoring high on measures of social dominance more arousing and more masculine.[50] One study found that women who were mated to dominant men with high facial masculinity (i.e., sexually dimorphic facial features) had more frequent and sooner orgasms during intercourse,[51] and this pattern of female orgasm has been associated with greater sperm retention.[52] While sexual dimorphism among humans is not as prominent as in species that are extremely polygamous,[53] men are generally taller and more robust, which women like.

Men also bring material investment to their mates and their children. Not surprisingly women across cultures tend to be attracted to material wealth in men (a highly consistent research finding).[54] In one study, researchers photographed a group of men in costumes representing differences in socioeconomic status—a Burger King® uniform and another costume comprised of a white shirt, designer tie, and a Rolex® watch. When the photos were presented, women stated that they were unwilling to date, have intercourse with, or marry the men in the low-status uniform, but were willing to consider all three types of relationships with men wearing the high-status one.[55] However, success in nature's interminable competition for resources is only as good to females as the winning male's willingness to share. Indeed, research has found that men's overall attractiveness is ultimately based on a combination of dominance and altruism, suggesting that women value dominant men who share.[56]

These patterns of attraction also extend to male gods. In Blackfoot Indian mythology, *Thunder* is a powerful male sky-being known for providing protection and provision (in the form of rain), for creating fearsome storms, and for striking men down and stealing their women. According to folklore, one young woman found all the mortal males in her company to be unsatisfactory. Knowing Thunder's strength and power, she sought him out, married him, and had his children. In doing so, she and her children gained rank in Blackfoot society—her children were thunderclaps, and, on behalf of Thunder, she brought the sacred pipe to the Blackfoot people. *Morning Star* was another male deity central to Blackfoot mythology. A woman became so enraptured with Morning Star's brilliance that she sought him out, married him, and had his child.[57]

For Christians, there is no more powerful male figure in the universe than God. The Spanish nun Saint Teresa of Ávila (1515–1582) offers perhaps the most illustrious example of the sexual attraction to a god of power. Below is an excerpt from Saint Teresa's diary:

> It pleased the Lord that I should sometimes see the following vision. I would see beside me, on my left hand, an angel in bodily form—a type of vision which I am not in the habit of seeing, except very rarely. It pleased the Lord that I should see the angel in the following way. He was not tall, but short, and very beautiful, his face so aflame that he appeared to be one of the highest types of angel who seem to be all afire. They must

be those who are called cherubim: they do not tell me their names but I am well aware that there is a great difference between certain angels and others, and between them and others still, of a kind that I could not possibly explain. In his hands I saw a long golden spear and at the end of the iron tip I seemed to see a point of fire. With this he seemed to pierce my heart several times so that it penetrated to my entrails. When he drew it out, I thought he was drawing them out with it and the pain left me completely afire with a great love for God. The pain was so sharp that it made me utter several moans; and so excessive was the sweetness caused me by this intense pain that one can never wish to lose it, nor will one's soul be content with anything less than God.[58]

It would be difficult to deny that Teresa was having a sexual experience with God. Another woman—Saint Thérèse of Lisieux or "Little Flower"—recounts a similar affair centuries later:

I do not know how to explain it; it was as if an invisible hand had plunged me wholly into fire. Oh what fire, and what sweetness at the same time! I was burning with love and I thought one minute, nay, one second more, and I shall not be able to support such ardor without dying. I understood then what the Saints have said of those states which they had experienced so often. For me I have but experienced it that once, only for an instant, and afterwards I fell back again into my habitual dryness. From the age of fourteen I have also experienced the assaults of love. Ah! how much I love God! But it was not at all to be compared to what I experienced after my offering to Love.[59]

Part of the Abrahamic God's appeal may be his power and dominance. We have explored how God, like powerful men, provides protection from outside tribes and from wild animals, and offers prowess in battle. But for God to have a sexual appeal to women, we would expect him to also be a great provisioner. The Abrahamic faiths abound with references to God's role as such:

- The Lord said to Moses, "Behold, I will rain bread from heaven" (Exod. 16:4);
- I will bless her with provisions. Her poor I will satisfy with food (Ps. 132:15);

- He hath given meat unto them that fear him: he will ever be mindful of his covenant. (Ps. 111:5);
- But my God shall supply all your need according to his riches in glory by Christ Jesus. (Phil. 4:19);
- Therefore take no thought, saying, What shall we eat? or, What shall we drink? or, Wherewithal shall we be clothed? . . . For your heavenly Father knoweth that ye have need of all these things. (Matt. 6:31–32);
- (They worship Him) so that God will reward their best deeds and give them more through His favors. God gives sustenance to whomever He wants without account (Koran 24:38);
- It is He who made the earth tame for you—so walk among its slopes and eat of His provision. (Koran 67:15);
- And it is He who feeds me and gives me drink. (Koran 26:79)

Combined, God's power and willingness to provide would seem a perfect match for women programmed with the tendency to prefer these traits in men. However, as with dominant male primates, male dominance strategies come with a significant cost to females, above and beyond sexual control.

The Cost to Women and Children

Between males and females we see a mutual exchange of resources— such as sex, childrearing, provisioning, and protection— that ultimately serves the reproductive interests of both sexes. However, the arms race of evolution can also lead to sexual inequities that result in high costs for females and their offspring. In women, these costs may become particularly large when the reproductive strategies of men, by their own evolutionary success, become extreme or exaggerated, or impart unchecked power. The following sections examine examples that have arisen where religion has ideologically championed a male reproductive strategy, often resulting in violent oppression of women across the globe.

Veiling

The Bible is especially frank in claiming that God wants women to submit to men, for example: "Wives, submit yourselves unto your own husbands, as it is fit in the Lord." (Col. 3:18). When the Apostle Paul wrote to the Corinthians advising them that their church was practicing improper worship, he argued that women are made to serve man, and man to serve God, therefore women are to pray to God in a manner that serves man—in chaste submission. Hence women are punished for failing to veil themselves in prayer:

> For a man indeed ought not to cover his head, forasmuch as he is in the image and glory of God; but the woman is the glory of man. For the man is not of the women: but the woman of the man. Neither was the man created for the woman: but the woman for the man. (1 Cor. 11:7–9)

Veiling women is a common religious practice that makes women less sexually tempting to rival males. Paul admonishes that if women dare to pray to God with their heads uncovered, they are to have their heads shorn. Hair has special value for men. Long hair in particular is regarded across cultures as a highlight of female beauty. Like other aspects of female beauty, the ability to grow long hair is also an evolutionary marker of physical health and fertility. Thus in Corinthians, this universal marker of female beauty is forcibly sheared as reprimand for the woman's arrogance. Men with God's backing will not suffer outward displays of beauty that could be interpreted as sexual availability.

Men in some sects of Islam have taken veiling to the absolute limit with the burka, a garment that conceals every inch of a Muslim woman's body. In this manner, men can manage temptation among rival males, and ensure no threat to their women's fidelity arises to damage their male honor. However, the dedication to hiding female sexuality can be taken to stupefying extremes. For instance, in 2002 a fire broke out at school in the holy city of Mecca. Religious police known as the *Commission for the Promotion of Virtue and Prevention of Vice* stopped firemen who tried to rescue schoolgirls trapped inside the blazing building. Some police were seen beating the girls to prevent them from escaping the inferno simply because they were not covered by the traditional black robes required by the faith. Fifteen teenage girls died in the fire because men promoting "virtue" prevented their escape.[60] Later that same year, riots broke out in Nigeria over the Miss World pageant that was to be held in the nation's capital, Abuja. From the start, the notion of women wearing bikinis incensed Nigerian Muslim men (and reportedly God) who rule a society in which female sexuality is under male jurisdiction. When a news reporter stated that Mohammed would have chosen one of the contestants as his wife, an explosion of violence tore across the region. In the end, two hundred people lay dead and hundreds more were injured. Many were stabbed, bludgeoned to death, or burned alive. Murderous rioters were heard yelling "God is great" and "Miss World is sin!"[61]

Violence against Women

Like their male primate cousins, men tend to batter women out of sexual jealousy. Research worldwide shows that sexual jealousy tends to be the leading motivation for spousal abuse, which is almost always perpetrated from man to woman.[62] Sexual jealousy is also a leading cause of spousal homicide around the globe, which again is almost always committed by jealous husbands rather than jealous wives. Further, because dominant men usually make the laws, they tend to skew them in favor of their evolutionary concerns. David Buss points out that men around the world, including in the United States, have been exonerated for killing their spouses if the murder was committed in response to the wife's adultery. In Texas, for example, such an act was considered something a "reasonable man" would do, and therefore it was legal to kill one's wife or her lover if caught in the act until as recently as 1974 when the law was changed.[63]

A casual read of religious canon uncovers a litany of violent acts against women. These are not aimless acts, but rather acts reflecting the reproductive strategies that God and man inherited from their primate predecessors. Jane Goodall recalls a dominant chimpanzee's response to sexual infidelity: "[W]hen Figan did notice he would race towards the pair and, very often, bash the female for her faithlessness."[64] Here we see the alpha male Figan bypass the male philanderers altogether and proceed to bash the female. The Bible has men doing the same.

Here is a verse from Deuteronomy that dumbfounds the rational mind and demonstrates how the exaggerated evolutionary advantage of men can be codified into religious law: "If a man happens to meet in a town a virgin pledged to be married and he sleeps with her, you shall take both of them to the gate of that town and stone them to death—the young woman because she was in a town and did not scream for help (22:23–24). In another startling verse, God goes on a rampage against his wives Samaria and Jerusalem (represented as the sisters Oholah and Oholibah) for infidelity:

> And I will direct my jealousy against you, that they may deal with you in fury. They shall cut off your nose and your ears, and your survivors shall fall by the sword. They shall seize your sons and your daughters, and your survivors shall be devoured by fire. They shall also strip you of your clothes and take away your beautiful jewels. Thus I will put an end to your lewdness and your whoring begun in the land of Egypt, so that you shall not lift up your eyes to them or remember Egypt anymore. . . . Your lewdness and promiscuity have brought this on you, because you lusted after the nations and defiled yourself with their idols. (Ezek. 23:25–30)

Here God advocates mutilation, stripping, and rape—behavior of a jealous, deranged man bent on extracting violent revenge for sexual indiscretion. For good measure, God had his wives' children burned alive.

The misogynistic brutality depicted in the Old Testament is recapitulated today in a practice known as "honor killing," which is now mostly concentrated in Muslim countries. These so-called honor killings—rooted in the notion that a woman's chastity is the property of her family—are perpetrated for various reasons concerning sexual control such as: real or rumored premarital loss of virginity, extramarital affairs, refusing a prearranged marriage, speaking to unrelated men, or even being raped. While some may argue that these murders are not religious in nature because honor killing is not explicitly prescribed in the Koran, it is clear that Islamic ethics, which place an extremely high value on female chastity, are often cited as the justification for the killings by the perpetrators. Islamic courts in many countries are notoriously lenient when it comes to punishing such murders, and many rule them to be justified. Horrors such as this are undoubtedly influenced by misogynistic scripture. For instance, in the Koran, God suggests beating unruly females:

> Men have authority over women because God has made the one superior to the other, and because they spend their wealth to maintain them. Good women are obedient. They guard their unseen parts because God has guarded them. As for those from whom you fear disobedience, admonish them and send them to beds apart and beat them. Then if they obey you, take no further action against them. Surely God is high, supreme. (Koran 4:34)

More specifically, however, murder is condoned by Sharia law. Sharia law is a set of religious laws inspired by the Koran and the examples set by Mohammed in the Sunnah—life guidelines for Muslims based on interpretations of the Koran. Today, a minority of nations enforces Sharia law, but Islamist movements around the globe are striving for formal reinstatement. Sharia law e12.8 prescribes killing for adulterers who are "unworthy (those who may be killed) includes . . . convicted married adulterers . . ." While the wording may seem to include equal penalties for both men and women, men are almost never killed for this infraction.

Infanticide in Men and God

As we have seen, male apes and monkeys will engage in infanticide, and this chilling behavior can confer a reproductive advantage. As beings programmed to nurture, cherish, and protect our children, the biological origin of infanticide is a hard pill for humans to swallow. Nevertheless, the empirical literature reveals that infanticide in humans follows the lines of inclusive fitness for males, just as it does for other primates. The number of child homicides committed by stepfathers or boyfriends is significantly higher, in some cases one hundred times higher, than those committed by biological fathers.[65] Further, differential abuse of stepchildren, particularly lethal beatings, is a male tendency found to generalize across Canada, Great Britain, and the United States, and in other parts of the world such as Australia, Korea, Malaysia, and Trinidad.[66] As with nonhuman primates, infanticide is exceedingly rare among human females. If biology were not a factor, we would not expect such clear prevalence across cultures, geographic regions, and primate species.

Infanticide, this pitiless animal behavior, is rather frequent in the Bible. In the book of Numbers, God advises not only to kill women who might possibly contain the genes of rival males, but also to kill all their male offspring. Virgins are to be kept alive and used for sex:

> Now therefore kill every male among the little ones, and kill every woman who has known man lying with him. But all the young girls who have not known man lying with him, keep alive for yourselves. (Num. 31:17–18)

In a rather infamous act of retribution against the pharaohs of Egypt, as a punishment for not letting Moses's tribe leave, God commits mass infanticide:

> Every firstborn son in Egypt will die, from the firstborn son of Pharaoh, who sits on the throne, to the firstborn son of the female slave, who is at her handmill, and all the firstborn of the cattle as well. There will be loud wailing throughout Egypt—worse than there has ever been or ever will be again. (Exod. 11:5–6)

And Babylon, yet again, is severely punished. We have heard of her harlotry before, and of God's response, but sexual infidelity has a tendency to produce rival offspring. The behavior described here mimics the technique used by the chimpanzees in the example above:

> Daughter Babylon, doomed to destruction, happy is the one who repays you according to what you have done to us. Happy is the one who seizes your infants and dashes them against the rocks. (Ps.137:8–9)

Finally, we see a prophecy against Babylon, where the pattern of primate infanticide and mating with the grieving mothers is captured in verse:

> Therefore I will make the heavens tremble; and the earth will shake from its place at the wrath of the Lord Almighty, in the day of his burning anger. . . . Whoever is captured will be thrust through; all who are caught will fall by the sword. Their infants will be dashed to pieces

before their eyes; their houses will be looted and their wives violated. (Is. 13.13–16).

A CASE STUDY

The violent acquisition of territory. Emasculation of male rivals. Sexual acquisitiveness. Winning exclusive sex through violence and cloistering. The sexualization of God. It is one thing to observe these phenomena in the Bible or Koran as ancient texts, but how relevant are they really in religious life? To answer this question, I offer a case study that draws on source material collected by historian Ramón A. Gutiérrez in his book, *When Jesus Came, the Corn Mothers Went Away*.[67] Here we see how the Spanish conquest of the Western Hemisphere provides an illustrative example of the alpha-god parable, played out in epic and dramatic form. In particular, we see religion's dominant-male ethos as wielded by men in violent competition.

When the Spanish rode into North America, clanging in their metal armor, they were a terrifying spectacle—the bearded men, the horses, the firearms, the dogs of war, all previously unknown to the native inhabitants, all predestined to mystify. Capitalizing on the awe, the Spaniards' first objective was to claim Indian territory in the name of Christ and God. They summarily demolished every religious building, shrine, and relic, and replaced every trace of Indian religion with crucifixes, reliefs of saints, and other Catholic iconography. The Spaniards understood the power of religious symbolism, and they used symbols of their Christian god as territorial markers of their dominant male. With these markers they scored out territories across vast spans of the Americas. After deposing the old deities with symbols of the new God, the Spanish went on to persecute and kill powerful religious medicine men, all with a great deal of purpose and forethought; religious conquest was the shortest route to domination. If religion were not important to this endeavor, the Spanish would have been unconcerned with the existing religious structure of the native peoples. But there was immense psychological utility in marking territories with symbols of a god with a reputation for being fierce, dominant, and male. Toward this end, they also forced the Indians to call themselves by Christian names.

Sexual domination was an indispensable charge of the conquest, and one that links it with conquests across primate history. The atrocities of Juan de Oñate, a Spanish explorer and governor of the New Spain province of New Mexico, are an infamous example. Upon arrival, Oñate quickly made a point to neutralize the Pueblo Indians' most sexually capable male rivals. Not long after riding into Acoma Pueblo, Oñate rounded up all Indian men over twenty-five years of age and hacked off one foot from each of them—in the name of Christ, the Church, and the Spanish crown. Any resistors were killed on the spot. If this were simply a "neutered" domination of a people, one would assume the mutilations would have been meted out more judiciously across sex and age—to include women and prepubescent boys. Further following Darwinian imperatives, Oñate's men summarily took to raping the Puebloan women as their entitlements of war.

From their own seats of dominance, similar behavior arose among the Franciscan friars, who were explicitly charged by the Church with orchestrating the spiritual conquest of the native populations across the region. Again, an important first step to any conquest is toppling the existing male-rank structures. Accordingly, a primary task of the Franciscans was to break alliances between young men and their fathers. This way fathers would drop in rank and sons could more easily acquiesce to the new dominant males—God, Christ, and the Friars themselves. Inspired by passages from the Bible, the Franciscans publicly humiliated native fathers or physically assaulted them when humiliation alone failed to recruit their sons. More notable, however, was the sexual nature of the Franciscans' play for dominance. Gutiérrez notes one common domination strategy practiced of the friars whereby they seized Indian men by their penises:

> The humiliation of fathers before their own children was most demeaning when the [Franciscan] fathers emasculated the men, thereby symbolically transforming them into women. A clerical technique occasionally used to render an obdurate and cocksure Indian submissive was to grab him by the testicles and to twist them until the man collapsed in pain. Pedro Acomilla of Taos complained in 1638 that Fray Nicholas Hidalgo "twisted [his penis] so much that it broke in half," leaving Pedro without "what is called the head of the member."[68]

Gutiérrez captures how common this was, citing numerous historical documents that describe friars ripping off the penises of Indian men, literally emasculating them and rendering them impotent sexual rivals. This was performed for a variety of reasons, from failing to submit to Christ to maintaining forbidden Indian customs. Here the alpha god paradigm is agonizingly clear: Indian men were to sexually submit to Christ, the dominant-male archetype, and the Franciscans exercised extreme brutality to accomplish such subservience, including attacking genitalia in the style of male apes and monkeys.

Like god-kings, the Franciscans were seen as possessing supernatural powers, and as bridges to God they were thus revered in a godlike manner by the Indians. This power differential was codified by the Catholic Church, which ceded absolute dominion over the Indians to the Franciscan order. The Indians were regarded as children and wards of the Franciscans or, in some cases, as animals. Consequently, Indians were often indentured by the Franciscans or flatly made slaves.

By dominating the Puebloan men, the Franciscans gained sexual access to Puebloan women. There is rich documentation of this, particularly from notes logged by inquisitorial courts. Gutiérrez relays numerous accounts, many involving violence. One example, among many, involved a friar named Nicolas Hidalgo who served the Taos Pueblo. An Indian woman stepped forward with a "half-breed" child allegedly belonging to Fray Nicolas and reported that Hidalgo had strangled her husband and then had sex with her, producing the child. Another Friar, Luis Martinez, was accused of "forcibly [raping] a woman, splitting her throat, and burying her in his cell {priestly quarters}."[69] The civil authorities turned away from prosecuting this rape and murder for fear of an Indian uprising.

To be fair, it was not uncommon for Puebloan women to give themselves to holy men within pre-conquest Indian communities. After the coming of the Spaniards, this tradition transferred over to the friars, many of whom were given sex willingly. The end result was that the seemingly unlikely Franciscan friars achieved differential reproductive success. One document has a friar reporting that "all the pueblos are full of the friar's children" and that many of the friars kept concubines.[70] Another Franciscan priest finally gave evidence against one of his brothers, relaying a litany of sexual conquests to his superiors:

María Encarnación and María de García, first cousins of María Guadalupe Valdez, who bore a daughter, Manuela Trujillo, María Lain, who bore a son, Ana María Fresquis, Rosa Mestas, who bore a son, La Roma, La Lupe Sánchez, Antonia Gallego, Ingacia Peña, the daughter of Isidro Medina who bore a son, the daughter of Alejandro Márquez, who bore a son, Manuel Trujillo, alias la Malinche, Soledad Tenorio.[71]

In this small historical glimpse, we can easily see male evolutionary paradigms grinding their way across the *Conquista* within a strongly religious context—the acquisition of territory, the sexual repression of males, the acquisition of females, and differential reproduction among despotic men.

Further, following the charge of monkeys and apes, the Franciscans hoarded Indian women, ostensibly for the sake of God. Cloisters were common among the Franciscans in New Mexico, as well as in other locations across America, where claims to female sexuality were secured by God's walled fortresses. The Franciscans spent a great deal of energy running off Indian male suitors and punishing them for trying to breach the cloister walls. While the stated reason for cloistering did not acknowledge any evolutionary motives, the behaviors played out in ways stereotyped across male primates of ages.

Pre-conquest Puebloan religion also saw women engaging in symbolic sex with their native gods. The Franciscans viewed the native religious practices—particularly those involving sex—as filthy, shameful abominations and made eradicating such practices from Indian culture a primary concern. Ironically, the Bride of Christ doctrine was embraced by the Franciscans and used to sway the hearts of the Indians. With it, both native men and women could be urged to submit themselves sexually to Christ in place of their deposed native gods. Again, the soul was made female so that both men and women could sexually serve the passions of Christ as his brides. Here is an extended passage from a Franciscan hymn with Gutiérrez's commentary included:

Listen as the bride (soul) cries out to her bridegroom (Christ): "enflame me and embrace me totally with the fire of your love so that my entire soul melts onto you, flows on you, and is united perfectly with you." Passionately she begs, "that He kiss me with a kiss of His mouth" with that kiss symbolizing their perfect union. When the Bridegroom beholds the soul's nakedness, "he pities her and spreads his cloak to cover her, saying that this is the moment reserved for lovers and for their sweet breasts, and he gives her wine to drink ... from the cellar of divine love." Now she wishes only that He "penetrate her intimately and to the depth of her heart. She wishes that He whom she desires would not show Himself to her under the exterior form, but as an intimate infusion into her, not that He appear to her but that He penetrate her." Pierced by the "lovers arrows," carried away on the "wings of love," enflamed by the "fires of love," she sings libations of praise, benedictions, and adorations to her spouse. Rapturously she utters, "O my Love! Ah, Love of my life!"[72]

The hymn was an allegory of human love, and was intended somehow to be chaste. Naturally, it often wasn't taken as such. After successively repeating the hymn, male Indians would report waking with homoerotic dreams of Christ. Documents show that for this they were urged to give themselves fifteen lashes and "symbolically," "mystically" *enter* Christ through his wounds:

> Place the eyes of our soul on the holiest wounds of your Redeemer and lord, hugging the cross, looking particularly at the holiest open wound and the blood which flows from him, from his entire person, body and cross which he shed for us; supplicating with loving affection that he defend you from your enemies.[73]

Gutiérrez notes that the wounds of Christ were often portrayed as disembodied in reliquaries and intentionally not in horizontal form but rather transformed vertically to appear like a vagina and pubic hair.[74]

The Franciscan's theocratic reign was heavily endured by the Puebloan people. At times the native people attempted to revert back to their own religions, invoking their native gods. For these indiscretions they were punished with a jealous, brutal fury. Whippings were common, and sometimes the Indians were even beaten to death. One such incident (again, among many) took place in 1655, when Fray Salvador de Guerra found that Juan Cuna, a Hopi Indian, was "worshipping idols." The priest whipped Cuna until the Indian was "bathed in blood." The next day, in the church itself, de Guerra gave Cuna another violent thrashing, eventually killing him in a shower of flaming turpentine. De

Guerra justified his actions to Church authorities as reasonable means to stamp out idolatry.[75] Torture by burning was also not uncommon.

After a time the Puebloans became tired of Spanish maltreatment, particularly as was delivered to them by the Franciscan priests. In the early 1670s, efforts to revive their forbidden practices gained momentum, and the Indians soon began revolting against the vicious religious tyranny of the Franciscans. It is striking how warfare between the Puebloans and the Spaniards was not only a war between men (or between cultures) but a war of rival gods. In 1672, the Abó Puebloans revolted and burned down the local church. Recalling their prior treatment, the Abó stripped the church's friar, Pedro de Ávila y Ayala, and flogged him mercilessly. After this they killed him and hung his naked body from a cross. As another symbolic gesture, the Indians slit the throats of three lambs and placed their bloody carcasses at the foot of the cross. The governor of the region, Juan Francisco Treviño, addressed the Indian sedition by launching a violent campaign against *idolatry*. He hung several known medicine men, whom he called "sorcerers." Forty-seven other medicine men were arrested, beaten, and then sold into slavery.[76]

Later, Popé, a medicine man from San Juan, began organizing a massive rebellion. Popé keenly pitted gods against God, entreating his followers to revive the alliance of traditional native gods. Under Spanish rule, only Christ was allowed to have multiple wives; elders and chiefs were burned or had their hair shorn as punishment for polygamy. In hopes of bringing differential reproduction back to men of his own culture, Popé promised that "[he] who shall kill a Spaniard will get an Indian woman, for a wife, and he who kills four will get four women, and he who kills ten will have a like number of women."[77]

Popé's forces eventually revolted and overtook Santa Fe. There is record of them shouting from the streets that "now the God of the Spaniards, who was their father, is dead, and Santa Maria, who was their mother, and the saints . . . were pieces of rotten wood" and that their own deities had never died.[78] After retaliating for the years of subjugation and abuse, the Indian forces of the region reclaimed their lands. Accordingly, they reinstated all their native shrines and religious symbols and tore down the crucifixes that mocked them on every corner. They also changed their Christian names back to Indian names and enacted punishments to anyone uttering the names Jesus or Mary.

American psychologist and philosopher William James makes the point that because the evolutionary mechanisms of our brains operate so efficiently, we are often unaware of their presence. For many believers, sexually submitting to a dominant male god, or even following his lead in winning sexual dominion, are behaviors so emotionally familiar that they may be succumbed to without question. This, however, is what makes them dangerous. If, as we learned in chapter 2, we *make the natural seem strange*, we take a first step toward questioning the gendered dictates of religious custom. Perhaps by questioning we may advance toward a spiritual ethics that does more to foster mutual well-being and less to repeat the violence and rapacity of our evolutionary past.

Chapter 5

COOPERATIVE KILLING, IN-GROUP IDENTITY, AND GOD

EVIDENCE IN THE MICROCOSM

Cooperation is a fundamental quality of life on earth. Symbiosis provides a principle framework for biologic complexity, and there is evidence that some parts of cells were once free-living parasitic bacteria that managed to form mutualistic relationships with their hosts, ultimately merging into singular, cohesive units of life. Accordingly, unicellular organisms cooperate to form tissues, tissues to form organs, organs to form organ systems, and organ systems to form complex multicellular organisms, such as humans. Within our bodies there are trillions of organisms cooperating in vast networks that comprise living, breathing, thinking human beings. In this sense then, we are literally constructed of many levels of cooperation.

Just as cooperation is a fundamental quality of life on earth, demonstrable in life's most basic forms, so is enmity. Viruses, for example, are simple chemical compositions that, with no cell structure or metabolism, seem to teeter on the edge between life and nonlife; some are as basic as a single strand of RNA covered by a protective protein coating. Because viruses have no metabolism, they require host cells to survive and perform their replicative functions, which they do at the expense of the host cell's survival—essentially robbing that cell of energy. Most viruses eventually kill their host cells, often by exploding them (a process known as *lysis*). We find hostility even in life-forms more basic than a single cell.

In nature, cooperation and hostility often come together. In a world in which even the most minute life-forms are designed to kill, joining forces with other organisms provides a clear survival advantage. As such, we find cooperative warfare in the microbiologic world. The strategic complexities employed in microscopic warfare draw striking comparisons to their human analogues and reflect adaptations that speak to the fierce pressures of selection at life's most basic level of organization. Here one scholar enumerates the similarities:

> ... military alliances (could apply to synergistic pathologies, where more than one pathogen act in concert) ... suicide mission (cells that self-destruct to kill the intruder), suicide bags (name applied to lysosomes that break open and release their contents destroying the cell ...) ... camouflage (coating on gram negative bacteria that inhibits recognition as foreign body by failing to provide earmarks of enemy) ... wolf in sheep's clothing (could be applied to viruses which have envelope made from host cell membrane) ... Trojan horses (bacteria which invade macrophages meant to destroy them and travel to other sites of the body protected from attack) ... distress signals (chemicals released by injured and dying cells) ... sabotage of communications (microbes commonly bind to cell signaling receptors on surface distorting or blocking communication ...) ... The key to a host's defense is being able to recognize its own cells and molecules from those of the pathogen (i.e. SELF from NON-SELF). In the military context, such recognition is accomplished by wearing different uniforms.[1]

On the human battlefield, the stakes are the same as in the microscopic wars waged inside us—survival and reproduction—and both theatres are ultimately administered by our genes. The analogy highlights another important point: a fundamental task of cooperative hostility is to recognize who is part of the cooperative alliance and who is against it.

For this task, humans have evolved a psychology to navigate interactions with both the in-group (a social group to which an individual belongs) and the out-group (a social group to which an individual does not belong). We have also developed cultures and religions as means to regulate in-group–out-group processes. Like secular culture, religions can engender great empathy and collaboration, but they can also bring about remorseless killing. This pattern of in-group altruism and out-group enmity has been termed *parochial altruism.*[2]

Religious enmity ultimately stems from the genetic strategies in which religion's out-group biases are rooted. Believers driven by religious hatred are for the most part unaware of the evolutionary designs of their behavior. Rather, they are often swayed by religious ideology that is modeled on their evolutionary psychology, which makes their stances and actions feel natural. To understand how this can come about, our task yet again is to make the natural seem strange.

ESTABLISHING BOUNDARIES WITH KIN ALTRUISM

In-group, Out-group

To begin, we revisit the idea of kin altruism. As a rule, cooperation is highly correlated with the amount of genetic material shared between individuals. Accordingly, kin altruism is a cornerstone of biological interaction, including that which occurs between humans. With brains shaped by eons of kin selection, the rules of kin altruism suffuse human thought, language, and behavior. Humans, with their unprecedented capacity for abstract thought and language, have even adopted the rules of kin altruism to influence each other—humans often symbolically exaggerate genetic relatedness in order to foster in-group cohesion and loyalty among nonkin. For example, fraternity members see themselves as "brothers," and soldiers see themselves as "brothers-in-arms." Similarly, countrymen share a common fatherland or motherland. Nonkin in-groups don't always require notional relatedness, but this strategy tends to strengthen affiliation.

As with in-group loyalty, this tendency to create fictive kin is pronounced in religion—for instance, God is called Our Father, his priestly representatives are addressed as fathers (or mothers), religious coadherents are brothers and sisters, and, collectively, the pious share a common ancestry as God's children. These designations tender a great deal of trust, compassion, and kindness between members of a shared faith. Creating notional brothers and sisters ends up being a powerful means of enticing people to close ranks.[3]

Contrarily, exaggerating genetic *differences* between others is a highly effective means of fomenting enmity. A common ideological strategy used in warfare is to regard the enemy as nonhuman; enemies are often seen as dogs, pigs, monkeys, monsters, or devils. Most people rallied to war by these images are hardly aware that their infrahumanizing rage is based on the logic of selfish genes.

Army psychologist David Grossman argues that humans are naturally disinclined to kill members of their own species. The "psychological cost of killing," as Grossman describes it, is both pervasive and measurable in warfare and in its aftermath. A disinclination to kill one's own species, he argues, is characteristic of many animals that engage in ritualized combat. Piranhas, as one example, are equipped with razor sharp teeth designed to tear other animals to shreds, but these fish notably don't do this to *one another*. The contests between them are symbolic, and rarely result in serious injury. The weaker fish often submits, and the stronger fish is likely to pull back when a competitor yields. These behaviors allow competitors to avoid life-threatening injuries and suggest that animals (directed by genetic strategies) are able to exert an emotional brake on violence between members of their own species. Humans come equipped with similar brakes, which—considering our ability to cooperate—must be very powerful. When humans engage in lethal combat, Grossman argues, they disengage those brakes in part by using strategic terms designed to make the enemy seem less human (i.e., less genetically related). This is what we have been calling *infrahumanization*, and in Darwinian terms is akin to saying, "You share no genes in common with me, therefore I owe no loyalty to you, and I can kill you like an animal."[4]

Philosopher David Livingstone Smith has compiled an extensive inventory of infrahumanization across the history of war.[5] He begins with an observation made by English philosopher David Hume, who, in 1740, spoke to our tendency to dehumanize the enemy and also to our extreme bias toward actions of the in-group:

> When our own nation is at war with any other we detest them under the character of cruel, perfidious, unjust and violent: But always esteem ourselves and allies as equitable, moderate and merciful. If the general of our enemies be successful, 'tis with difficulty we allow him the figure and character of a man.[6]

Smith goes on to cite early anthropological observations of aboriginal people who saw their enemies as nonhuman animals, which made them much easier to hunt and kill as prey. The propaganda of war offers another revealing look, which Smith argues taps an ancient fear of predation by outside species:

> A Union poster from the American Civil War shows a heroic club-wielding General Scott of the Union army. He is poised to bludgeon a gigantic, nine-headed serpent. Seven of the monster's heads are those of leaders of the Confederacy.
>
> An American cartoon from the Spanish American War represents Cuba as a huge, sinister ape-man, complete with protruding fangs, holding a bloody knife and hulking over the grave of U.S. Servicemen killed in the battleship *Maine*, which blew up in Havana Harbor in 1898.
>
> A Taiwanese cartoon depicts a hapless man in a wooden boat about to be devoured by an enormous shark. The boat is labeled "Taiwan" and the shark is labeled "China."
>
> A Soviet poster from the 1950s shows a wolf dressed in a suit and tie, removing a mask from its grotesque, snarling face. The mask bears the countenance of the United States Secretary of State Dean Acheson.[7]

Similar out-group depictions, particularly during warfare, could be exemplified indefinitely. Thus, considering that people are so often prone to seeing outsiders as nonhuman while calling themselves *the people, the human beings, the chosen people, the righteous people,* or even *people at the center of civilization,* it should not surprise us that religious canon also embodies these contrived distancing strategies. This is seen across the Abrahamic religions. Here is an example from the Koran in which Allah, so disgusted with the Jews, transforms them into apes and pigs:

> Those who incurrent the curse of Allah and his wrath, those of whom some he transformed into apes and swine, those who worshipped evil—these are [many times]worse in rank, and far more stray from the even path. (Koran 5:60)
>
> And well ye know those amongst you who transgressed in the matter of the Sabbath. We said to them, "be ye apes, despised and rejected." (Koran 2:65)

Clearly, Islamic doctrine includes infrahumanization, and it is not hard to understand how characterizations such as these could stoke the ancient coals of out-group mistrust, ultimately enflaming violence against Jews. With the main texts of Judaism predating Islam by hundreds of years, we would not find infrahumanization specifically directed toward Islam in texts such as the Old Testament or the Talmud—but finding the sons of Abraham infrahumanizing others in the modern age is not difficult. For example, when Israeli soldiers assault and torture Palestine children, they do so regarding them as "dogs."[8]

For a case study of infrahumanization in religious violence among Christians, we look to the Thirty Years' War (1618–1648). During this period Europe was ripped apart by bloody conflict, not between Christianity and Islam or Christianity and Judaism, but among Christians themselves; millions of Protestants and Catholics battled explicitly over theological differences of opinion. The divisions were not only between Catholics and Protestants; Protestants were divided into warring interests amongst themselves, with the Calvinists, Lutherans, Anabaptists, and Unitarians each fighting each other for religious supremacy and, importantly, the territories that it afforded.

When the Lutherans defined their doctrinal standard in the Book of Concord in 1580, they expelled the Calvinists from their territory in Germany. The Calvinists then drew up their own standard, the Heidelberg Catechism, which incensed both Catholics and Lutherans. The Calvinists in turn suppressed the Unitarians and sentenced to death those who questioned Calvinist doctrine. The Lutherans and the Calvinists began killing dignitaries and other people for alleged Calvinist or Lutheran leanings, respectively. Infrahumanization propelled the hatred. For example, in 1582 Lutheran pastor Nivander published a paper outlining forty characteristics of wolves and likened each precisely to characteristics of Calvinists. This tactic spread like wildfire across the various religious fiefdoms and readied Lutherans to butcher Calvinists. The Lutherans assaulted the Catholics in a similar manner. Historian Will Durant notes that "Words like *dung, offal, ass, swine, whore, murderer* entered the terminology of theology," and that this sentiment was captured in the political art of the day—for example, in German woodcuts depicting the pope as a sow giving birth to Jesuit piglets.[9] Theologians argued vehemently over the minutia of religious edicts and practices and used infrahumanization to artificially widen ideological differences. One Lutheran pamphlet read, "If anybody wishes to be told, in a few words, concerning which articles of the faith we are fighting with the Calvinist brood of *vipers*, the answer is, all and every one of them . . . for they are no Christians, but only baptized Jews and Mohammedeans" (italics mine).[10]

Each opposing group in the Thirty Years' War took to slaughtering one another like the animals they portrayed. In Germany and Austria alone there were estimated to be some 7.5 million human lives destroyed.[11] Like all wars, the Thirty Years' War wrought immense human suffering, involving not only killing but large-scale rape, torture, starvation, and infanticide. It is worth emphasizing that these atrocities were Christian on Christian—groups of people worshipping the same god and the same Christ. In this case, exaggerating genetic differences by making fellow Christians into animals (as opposed to brothers) eased the way to their extermination.

While humans are unique in using religious ideology to bolster in-group–out-group boundaries, their ideological strategies are thickly veined with evolutionary significance. This is not to make the argument that infrahumanization *causes* war. Rather, it is an effective way to grease the gears of war in a manner that reanimates the emotions of our ancestors who rallied against outsiders for survival.

Because both recognizing kin and fearing outside threats are survival adaptations, it makes sense that infrahumanizing a perceived outside threat serves the function of terror management—doing so, and closing ranks within the tribe, must have afforded survival advantages to our ancestors. This intrinsic human tendency can be elicited in the research lab. Terror Management Theory (TMT) studies find that subjects will rate outsiders as having more animalistic traits and insiders, more human traits, when fears of death are artificially manipulated.[12] The research also finds that subjects with lower self-esteem tend to infrahumanize more when mortality fears are activated than those with higher self-esteem.

Infrahumanization can be considered an ideological means to foster in-group cohesion based on logic that is evolutionarily familiar to us—the logic of kin altruism. Scholars in evolutionary sciences have operationalized other forms of cooperation that also help to explain why humans are so successful at extending cooperation beyond their genetic relatives. These mechanisms bear strongly on the emergence of cooperative hostility.

RECIPROCAL ALTRUISM AND INDIRECT RECIPROCITY

For humans, the advantages of cooperation materialize not just in things like hunting, gathering, childrearing, or warfare, but in virtually every avenue of survival imaginable. It is fair to say that the survival of the species as we know it would have been impossible without the cooperative enterprise.

Recall that in kin selection, genes "design" brains that entice cooperation between genetically related individuals. In doing so, genes "ensure" that copies of themselves (residing in genetic relatives) are replicated. Reciprocal altruism, a concept developed by American evolutionary biologist Robert Trivers, explains how seemingly altruistic acts between genetically *unrelated* (or distantly related) individuals evolved. Trivers defined reciprocal altruism as "behavior that benefits another organism, not closely related, while being apparently detrimental to the organism performing the behavior, benefit and detriment being defined in terms of the contribution to inclusive fitness."[13]

As Trivers points out, the apparent detriment here often becomes a future benefit. Altruistic behaviors are offered with the expectation that help will later be reciprocated, thus conferring a survival advantage. The individual offering aid to another may not be acting to perpetuate shared genetic material in that moment, but is still relying on the expectation that—somewhere down the line—he or she will receive aid in return, which will benefit survival and reproduction. Reciprocal altruism strongly influences a wide variety of human interactions—from the human dyad to the world economy—and makes humans much more effective at solving adaptive problems, chiefly because it allows us to extend our networks beyond genetically related kin.

However, given the timeless propensity of genes and the people they program to be "self-interested," some individuals inevitably attempt to take advantage of others' altruism by seeking help without reciprocating. If successful, they increase their own fitness by being recipients of help without risking their fitness by providing it—for example, the hunter who has a personal barbecue out in the bush instead of bringing his quarry back to camp yet demands a split of his comrade's kills. Natural selection has responded by providing sensitive cheater-detection systems in species that reciprocate, but this in turn drives selection for individuals who are very good at cheating on the sly. From this dynamic arises another evolutionary arms race, with increasingly sly cheaters pushing the development of ever more sensitive cheater-detection systems and vice versa. In the end, people have become extremely crafty cheaters and are also very wary of cheaters in cooperative arrangements.

Indirect reciprocity is similar to reciprocal altruism, but can occur without the expectation of a reciprocal response directly from the person receiving help. This process relies on reputation. For instance, if an individual performs an act of altruism, and if the recipient informs others that he or she has been helped, then the helper accrues reputational capital. Having a reputation for helping pays off by making others more confident in cooperating with the helper— even those who have not directly observed the helping. Thus, indirect reciprocity can extend the reach of cooperation across even greater numbers, allowing us to build sprawling societies of cooperators capable of working toward common goals.

Across the various cooperative enterprises, cheater detection remains critical. The cost lies in investing time or resources in those with no intention of returning the favor. Clearly, in a world of limited resources (and where cheaters cheat), one would want to be cautious about being altruistic. We can all think of examples of those who would fake signals of commitment to gain trust, ultimately to take unfair advantage. Because deceit of this kind has historically impacted survival, humans continue to find it very threatening. For these reasons, humans expend great effort in cooperative undertakings to establish in-group membership, define codes of in-group reciprocity, and root out cheaters. One means of systematizing this process is by developing unique signals of commitment. Behavioral ecologist William Irons has argued that "for such signals of commitment to be successful they must be hard to fake. Other things being equal, the costlier the signal the less likely it is to be false."[14] In other words, when the cost of faking signals outweighs the benefit, then cooperation is on a safer footing.

Costly signals therefore allow in-group members to cooperate with some degree of certainty, and eschew expending time and energy trying to determine the trustworthiness of their neighbors. This idea is known as Costly Signaling Theory (CST), and religion is replete with examples.

Anthropologist Richard Sosis writes about religion's "three Bs"—*behavior, badges and bans*—as forms of costly signals.[15] Sosis's three Bs are easy to uncover. Many religions require behaviors that are difficult to fake, such as long pilgrimages and regular prostration, or time-consuming, such as praying the rosary or repeating religious verse. Similarly, religious life is filled with badges, such as ritualistic scarring, circumcision, long beards, curled locks of hair, burkas, et cetera. Finally, religions are filled with all sorts of bans—for example, restrictions on eating, sexual behavior, language use, or even substances such as coffee, tobacco, or alcohol.

Sosis offers the example of ultraorthodox Jews in Israel, known as *Haredim*—which means "[God] fearing or trembling ones":

> Women sport long sleeve shirts, head coverings or wigs (and occasionally both), and heavy skirts that scrape the ground. In their thick beards, long black coats, and black pants Haredi men spend their days fervently swaying and sweating as they sing praises to God in the desert sun. Many of them wear *striemels*, thick fur hats that were undoubtedly helpful in surviving the long and cold Eastern European winters where their ancestors had lived, but probably should have been left at the border when they immigrated to the Holy Land. By donning several layers of clothing and standing out in the mid-day desert sun, these men are signaling to others: "Hey! Look, I'm a *Haredi* Jew. If you are also a member of this group you can trust me because why else would I be dressed like this? Only a lunatic would spend their afternoon doing this *unless* they believed in the teachings of Ultra-Orthodox Judaism and were fully committed to its ideals and goals.[16]

Donning furs in the desert is a particularly difficult-to-fake signal of commitment. Precisely because of their difficulty, such behaviors create confidence in the sincerity of the signaler and allow trusting social networks to form around those willing to swelter together. Costly signaling eases the way for reciprocal altruism and indirect reciprocity, essentially demarcating who is safe to trust in cooperative exchanges and who is to be mistrusted, avoided, or sometimes even persecuted. The ubiquity of rituals qualifying as costly signals points to the fitness value of making in-group–out-group distinctions.

GOD AS WAR-MAKER

Patterns of Primate Alliance-Making

Like humans, the social lives of nonhuman primates are governed by rules of reciprocity. Rules of exchange in coalitional violence are clear. For instance, when calling for assistance during a fight, individuals are more likely to be aided by those they have previously helped in a fight (or groomed, or have shared food with).[17] Reciprocating help in fights is an important social rule that is also enforced with violence. For instance, chimpanzees who fail to reciprocate with help in a fight will often be attacked (rather than helped) when they call for backup.[18] Further, chimps will methodically isolate and attack group members who assisted their rivals in fights against them.[19] Some researchers have concluded that the single most common cause of aggression within primate societies is perceived violation of social rules.[20]

Rank status plays an important role in alliance-making. To begin, primates appear to have an obsession with those of high status. In the research lab, for instance, male macaque monkeys will choose to view images of high-status individuals (and female genitalia) over food, but will require food overpayment to view lower-status individuals.[21] Research shows that human toddlers also tend to prefer and emulate higher-status over lower-status individuals.[22]

The penchant for selectively focusing on high-status individuals may be traceable to the advantages of forming alliances with them—having "friends in high places" pays, particularly if you live in a world in which cooperative hostility is the norm. Accordingly, we find that primates spend disproportionate energy attempting to form alliances with higher-status individuals. For example, in the strict matrilineal hierarchies of vervet monkeys, macaques, and baboons, support in fights is often given to the higher-ranking female, who will later intervene in conflicts on behalf of the helper.[23]

Dominants also rely on alliances to assume or maintain their rank.[24] Further, high-ranking primates will police relationships in their subordinates—for instance, by punishing behaviors such as sharing food with or grooming forbidden individuals.[25] Frans de Waal has argued that this type of behavior is a strategy to interfere with possible alliance formation.[26] When interfering doesn't work, alliances are punished with violence.

Human primates also make these kinds of patterned alliances and reenact them in their relationships with their man-based gods. Humans seek alliances with their (presumably higher-ranking) gods in battle and will ritualize food sharing (e.g., in the form of offerings, sacrifices, etc.) in order to secure them. Like any dominant primate, God is said to help those who ally with him and punish those who ally with the competition.

Humans will go to great lengths to secure alliances with God in battle, including making extreme and costly signals. However, this behavior violates central tenets of the Abrahamic-god concept, mainly omnipotence and immortality. Costly signals in coalitional violence are the mark of biological entities who risk being killed. Mortal beings generally cannot afford to behave altruistically without the reassurance that help will be reciprocated. But God is an immortal being; conceivably he could offer his alliance in battle with no risk to his existence. The same is true for omnipotence. According to scripture there is no being more potent than God. He therefore could never be harmed, and any alliance gestures that humans may make to him are inconsequential. Nevertheless, to secure his alliance humans will go to great extremes to demonstrate and enforce their association with God, often in a manner that generates grave human suffering—both among the enemy and among those who would refuse allegiance to him. The logical explanation for the illogic of making costly signals to God is that humans use the rules of reciprocity of their own societies to guide their "interactions" with him. Though transformed and elaborated upon through culture, such rules are grounded in the social structures of our primate ancestors and enacted unconsciously through our evolved psychology.

Costly Signals with God for Help in Killing

Charles Darwin recognized that warfare could be a powerful evolutionary force shaping in-group cooperation among humans.[27] Indeed, although warfare would seem to be the very definition of human antagonism, it does require the most far-reaching, coordinated acts of costly cooperation across the vast repertoires of possible human group behaviors. Given the stakes involved, warfare requires immense trust in reciprocity, for relying on an unreliable comrade in war brings the ultimate price. Perhaps for this reason severe punishments are delivered to those who violate the rules of reciprocity in war—treason and cowardice, for example, have been punished not uncommonly with execution across the history of organized warfare. Accordingly, the alliances humans make with other humans in warfare are perhaps stronger than any other. It has been noted, for instance, that the bonds between fighting men can be stronger than the bonds between those men and their wives,[28] and so reliant are brothers-in-arms on each other that the primary motivation for fighting becomes neither country nor cause, but *one another*, and the preservation of one's honor in the eyes of one's comrades.[29]

With brains shaped by millions of years of cooperative exchanges, humans are primed for making costly signals to secure alliances. This cross-cultural tendency extends to our concepts of God, with humans expecting God to operate on the premise of Costly Signaling Theory (CST)—a human means of establishing loyalty.

The ancient Mayans offer an example. Their supreme god Tohil demanded blood in exchange for the gift of fire. Worshippers would perform bloodletting—men, by piercing their penises, and women, their tongues and ears. By dripping blood onto paper and setting it aflame, the dutiful fanned their burnt offerings up to the heavens.[30] The ancient Aztecs had similar arrangements with their gods. The supreme feathered serpent god Quetzalcoatl is believed to have created humans through an act of sacrifice—by piercing his own penis and sprinkling the world into existence with his blood. Dutiful Aztec lords would return the gesture by puncturing themselves about their bodies, penises included, with the spiny thorns of the maguey plant or filed bones.[31]

In a similar instance of genital mutilation, the first landmark act of costly signaling in Judeo-Christian creed is the covenant between

Abraham and God. Indeed this covenant, this costly signal, is foundational to the Judaic religion. As the story goes, God asked Abraham to cut off his foreskin (and the foreskins of all his male descendants and male slaves) to prove his loyalty. In exchange for this painful act, God ceded the territories of Israel to Abraham and all his descendants in perpetuity. Per God:

> You are to undergo circumcision, and it will be the sign of the covenant between me and you. For the generations to come every male among you who is eight days old must be circumcised, including those born in your household or bought with money from a foreigner—those who are not your offspring. Whether born in your household or bought with your money, they must be circumcised. My covenant in your flesh is to be an everlasting covenant. Any uncircumcised male, who has not been circumcised in the flesh, will be cut off from his people; he has broken my covenant. (Gen. 17:11–14)

Just as men of war form strong alliances with one another, they form alliances with their male god, projecting onto him their own reflection—for example, "The Lord is a man of war" (Exod. 15:3). Accordingly, God acts like a man of war and provides support in warfare. It should not be difficult to understand how primates would take courage from a battle alliance with the most powerful dominant male in the universe. Here are a couple examples of how this dynamic is described in the Bible:

> Blessed be the LORD my strength which teacheth my hands to war, and my fingers to fight. (Ps. 144:1)

> And the LORD said unto Moses, Fear him not: for I have delivered him into thy hand, and all his people. . . . So they smote him, and his sons, and all his people, until there was none left him alive: and they possessed his land. (Num. 21:34–35)

The Aztecs made similar costly signals to their god of war, Huitzilopochtli. Late in Aztec history, the powerful leader Tlacaelel began uniting disparate Aztec states through a series of military campaigns under Huitzilopochtli. Through those battles, Huitzilopochtli eventually ascended to become the most powerful male god of the Aztec pantheon. In exchange for success in war, the Aztecs provided food to Huitzilopochtli in the form of sacrificial blood from their conquered enemies. The practice became central to their religio-militaristic culture. In a conversation with the Aztec emperor Motecuhzoma, Tlacaelel discusses the plans for a grand temple for the supreme Aztec god of war:

> There shall be no lack of men to inaugurate the temple when it is finished. I have considered what later is to be done. And what is to be done later, it is best done now. Our god need not depend on the occasion of an affront to go to war. Rather, let a convenient market be sought where our god may go with his army to buy victims and people to eat as if he were to go to a nearby place to buy tortillas . . . whenever he wishes or feels like it. And may our people go to this place with their armies to buy with their blood, their hearts and lives, those precious stones, jade, and brilliant and wide plumes . . . for the service of the admirable Huitzilopochtli.[32]

Huitzilopochtli was also regarded as the god of the sun, and the blood nourishment was also offered in exchange for his brilliance, which kept this world in existence. Often the hearts of sacrificial victims, many of whom were prisoners of war, were extruded with a jade knife while the victim was still alive. Like the Aztecs, the Mayans also sacrificed their prisoners of war. Ancient reliefs depict them decapitating, scalping, burning, or disemboweling their victims.[33]

In a similar vein, Abraham showed his loyalty to God by his unquestioning willingness to sacrifice his son Isaac (Gen. 22:1–19). God reciprocates in decidedly evolutionary terms by offering Abraham alliance in war (and progeny) in exchange for his submission:

> Because you have obeyed me and have not withheld even your son, your only son, I swear by my own name that I will certainly bless you. I will multiply your descendants beyond number, like the stars in the sky and the sand on the seashore. Your descendants will conquer the cities of their enemies. (Gen. 22:16–19)

The blood sacrifice of another Old Testament figure, Jephthah, did not receive a last-minute reprieve. In exchange for alliance in warfare, Jephthah burned his daughter alive.

> And Jephthah made a vow to the LORD: "If you give the Ammonites into my hands, whatever comes out of the door of my house to meet me when I return in triumph from the Ammonites will be the LORD's, and I will sacrifice it as a burnt offering." Then Jephthah went over to fight the Ammonites, and the LORD gave them into his hands. He devastated twenty towns from Aroer to the vicinity of Minnith, as far as Abel Keramim. Thus Israel subdued Ammon. When Jephthah returned to his home in Mizpah, who should come out to meet him but his daughter, dancing to the sound of timbrels! She was an only child. Except for her he had neither son nor daughter . . . and he did to her as he had vowed. And she was a virgin. (Judg. 11: 30–34, 39)

So powerful is this idea of sacrificing one's child to signal alliance that it ultimately became the most fundamental premise of Christianity, which claims that God sacrificed his only begotten son to ensure his human followers would have an afterlife. As captured in verse:

> God so loved the world, that he gave his only Son, that whoever believes in him should not perish but have eternal life. (John 3:16)

Here the Christian God allows his own son to be killed, an act which draws absolute awe from his followers. But the act is only astonishing to beings bound by genetic replication, to those for whom kin altruism has important survival implications. Remember that God does not need to perpetuate his line by genetic means; he is omnipotent. He is portrayed in scripture as creating by his voice, or by the Holy Spirit, and is not subject to the process of natural selection. God has the power to beget a billion sons, not to have only *one* begotten son. Perhaps more to the point, God as described has the power to *support* a billion sons without requiring reciprocity from puny humans. Therefore such a costly signal from God is an empty gesture, practically speaking, but it is deeply symbolic to followers who quite literally live and die on the basis of cooperation.

Once again pointing to the centrality of cooperation, God, like his primate counterparts, makes it absolutely clear that he does not like his followers to make costly signals to other gods, suggesting this may be considered cheating on the reciprocal exchange—for example, "He who sacrifices to any god, except to the LORD only, he shall be utterly destroyed" (Exod. 22:20).

Also like other primates, God requires alliance against his rivals (other gods) and against those who would ally with those gods in turn. His followers have been known to enforce alliance with God by slaughtering his rivals' followers and offering their flesh as burnt offerings.

> Suppose you hear in one of the towns the LORD your God is giving you that some worthless rabble among you have led their fellow citizens astray by encouraging them to worship foreign gods. In such cases, you must examine the facts carefully. If you find it is true and can prove that such a detestable act has occurred among you, you must attack that town and completely destroy all its inhabitants, as well as all the livestock. Then you must pile all the plunder in the middle of the street and burn it. Put the entire town to the torch as a burnt offering to the LORD your God. That town must remain a ruin forever; it may never be rebuilt. Keep none of the plunder that has been set apart for destruction. Then the LORD will turn from his fierce anger and be merciful to you. He will have compassion on you and make you a great nation, just as he solemnly promised your ancestors. "The LORD your God will be merciful only if you obey him and keep all the commands I am giving you today, doing what is pleasing to him. (Deut. 13:13–19)

The God of Islam also brokers deals with men in the manner of costly signals. What follows is an example how in exchange for lives in battle he gives a commodity highly valued among desert peoples (water) and also admonishes to eschew cooperation with those who don't cooperate with him:

> Those that suffered persecution for My sake and fought and were slain: I shall forgive them their sins and admit them to gardens watered by running streams, as a reward from God: God holds the richest recompense. Do not be deceived by the fortunes of the unbelievers in this land. Their prosperity is brief. Hell shall be their home, a dismal resting place. (Koran 3:195–96)

The costly signal may also be literally a financial cost. In the Koran, God rewards men for using their riches to finance military expansion and for sacrificing their lives in combat by giving them either triumph or paradise.

> Those who believe, and have left their homes and striven with their wealth and their lives in Allah's way are of much greater worth in Allah's sight. These are they who are triumphant. (Koran 9:20)

> Allah hath purchased of the believers their persons and their goods; for theirs (in return) is the garden (of Paradise): they fight in his cause, and slay and are slain: a promise binding on Him in truth, through the Law, the Gospel, and the Koran: and who is more faithful to his covenant than Allah? Then rejoice in the bargain which ye have concluded: that is the achievement supreme. (Koran 9:111)

In exchange for fighting for his doctrine—"O you who believe! fight those of the unbelievers who are near to you and let them find in you hardness" (Koran 9:123)—God rewards with victory in battle, forgiveness, love, and Paradise:

> O ye who believe! Shall I show you a commerce that will save you from a painful doom? Ye should believe in Allah and His messenger, and should strive for the cause of Allah with your wealth and your lives. . . . He will forgive you your sins and bring you into Gardens underneath which rivers flow, and pleasant dwellings in Gardens of Eden. That is the supreme triumph. And (He will give you) another (blessing) which ye love: help from Allah and present victory. (Koran 61:10–13)

> Surely Allah loves those who fight in His way. (Koran 61:4)

> He it is who has sent His Messenger (Mohammed) with guidance and the religion of truth (Islam) to make it victorious over all religions even though the infidels may resist. (Koran 61:9)

In short, as a dominant male, the god of the Abrahamic religions is turned to for coalitional violence, much as men turn to other men. Across cultures, geography, and epochs we find stories men of seeking cooperation with gods in exchange for alliance in warfare, which speaks to an ancient primate legacy reenacted in religious violence.

THE GREAT OUT-GROUP PREJUDICE OF HUMANKIND

The act of taking another person's life with malice aforethought (as opposed to accidentally) has been considered the worst violation of all social rules since antiquity. Accordingly, we find prohibitions against killing within every society. This rule is so comprehensively enacted in humans, and so intuitive, that it appears to connote the moral consensus that humans fundamentally possess a right to not be killed by other humans. But while this ethic is seemingly universal, it is far from absolute—while strong proscriptions exist for killing in-group members, killing the out-group members has often been allowed, encouraged, or even obligated in every human society. Thus this seemingly universal injunction against killing ends up being one of the most selectively enforced taboos of human life. To be fair, this judicious hypocrisy has been a matter of survival with humans, much as it has with nonhumans; groups unwilling or unable to kill are typically either annihilated or subsumed by those who are.

Probably for these reasons, religions have adopted this logic with ease. For example, commandment number six of Judeo-Christian law, *thou shall not kill*—at least as it has been applied for centuries—really means *thou shall not kill members of your own community*. Religious scholar John Teehan nails this point[34] by reminding us of the first thing Moses did when he descended from Mount Sinai; bearing a tablet freshly engraved with "Thou shall not kill," Moses summarily began to slaughter all those who had committed sins while he was away:

> Then Moses stood in the gate of the camp, and said, "Who is on the Lord's side? Come to me." And the sons of Levi gathered themselves together to him. And he said to them, "Thus says the Lord God of Israel, 'Put every man his sword on his side. And go to and fro from gate to gate throughout the camp and slay every man his brother, and every man his companion, and every man his neighbor.'" And the sons of Levi did according to the word of Moses; and there fell on the people that day about three thousand men. And Moses said, "Today you have ordained yourselves in the service of the Lord." (Exod. 32:26–29).

The morality of the seminal sixth commandment is predicated on in-group logic. *Thou shall not kill* establishes a rule that killing members of the in-group is considered murder, whereas different principles apply to killing out-group members—such killings are often acceptable, righteous, and regularly sanctioned by God. There are endless examples of out-group killing in the Bible. For example, when the lands of Heshbon were delivered to the Hebrews by God, the Hebrews did not passively receive those spoils. They boast that they "utterly destroyed every city, men, women and children; we left none remaining" (Deut. 2:34). From there they rampaged into Bashan where they,

> smote him until no survivor was left to him. And we took all his cities at that time—there was not a city which we did not take from them—sixty cities. . . . And we utterly destroyed them, as we did to Sihon the king of Heshbon, destroying every city, men, women and children. (Deut. 3:4–6)

Teehan argues, rightly I believe, that these were not "weaknesses of humans perverting God's goodness and mercy, for these are all divinely ordered massacres."[35] By every definition of the word, the behaviors described above are genocide—committed at the explicit command of a dominant male god. In this way, in-group loyalty may be the moral blinder obstructing the eyes of the religious who commit horrors in the name of their faith while singing the praises of their own moral rectitude. Contrarily, out-group members are seen as devils, their religious practices, as witchcraft, and their cultures, as morally bankrupt—which makes killing them all the easier.

While parochial altruism strongly defines Judeo-Christian morality, it flourishes in Islam. First, the rules of parochial altruism are declared bluntly in the following verse: "Muhammad is the messenger of Allah. And those with him are hard against the disbelievers and merciful among themselves" (Koran 48:29). But there are also texts that seem to police every angle of this arrangement. For one, much as Judeo-Christians strive to create unity and notional siblinghood, the Koran (and Muslim culture) repeats notional brotherhood successively, for example, "The Believers are but a single Brotherhood: So make peace and reconciliation between your two (contending) brothers; and fear Allah, that ye may receive Mercy" (Koran 49:10); "And hold fast, all of you together to the rope of Allah, and do not separate" (Koran 3:103). With the alliance of insiders thus established, the next step is to foment distrust of outsiders, which the Koran also does at great length. For example:

Let the believers not make friends with infidels in preference to the faithful—he that does this has nothing to hope for from God—except in self-defense. (Koran 3:28)

Believers do not make friends with any but your own people. They [outsiders] will spare no pains to corrupt you. They desire nothing but your ruin. (Koran 3:118)

Believers do not seek the friendship of the infidels and those who were given the Book before you [i.e., Jews and Christians], who have made your religion a jest and a pastime. (Koran 5:57)

You see many among them making friends with unbelievers. Evil is that to which their souls prompt them. They have incurred the wrath of God and shall endure eternal torment. . . . You will find that the most implacable of men in their enmity to the faithful are the Jews and the pagans, and that the nearest in affection to them are those who say, "We are Christians." (Koran 5:80–82)

And the Jews say: Ezra is the son of Allah; and the Christians say: The Messiah is the son of Allah; these are the words of their mouths; they imitate the saying of those who disbelieved before; may Allah destroy them; how they are turned away! (Koran 9:30)

Now with the out-group defined as unbelievers, the dominant male God of Islam prescribes out-group enmity by advocating retribution and assault against them:

Such are those that are damned by their own sins. They shall drink scalding water and be sternly punished for their own belief. (Koran 6:70)

No sooner will their skins be consumed than We shall give them other skins, so that they may truly taste the scourge. God is mighty and wise. (Koran 4:56)

They shall sigh with remorse but shall never come out of the Fire. (Koran 2:167)

Slay them wherever you find them. (Koran 9:5)

As with the genocide described in Deuteronomy, burning someone alive repeatedly (if only by magically reanimating their cooked skin), forcing one's victim to drink scalding water, and ignoring pleas for mercy and forgiveness are all behaviors that if perpetrated on in-group members would be considered the morally incomprehensible and macabre acts of a psychopath, someone with unspeakable disregard for in-group rules.

But in the eyes of many religious followers, the very concept of morality is synonymous with adherence to religious edicts and practices, including those as vicious as the above. Further, from the perspective of a religious insider, religious morality often reflects a higher order of moral reasoning, an unalienable set of ethical precepts that are *principally* sound (and righteous, etc.). But, again, these morals do not survive much past the periphery of the religious in-group, a fact that would seem at odds with their imagined transcendence. This limitation plays a hugely important role in facilitating religious warfare. Recall that warfare—defined as cooperative killing of *out-group* members—is by definition collectively sanctioned by the in-group members engaging in it.

THE SOCIOPATHY OF THE IN-GROUP

Some crimes, such as rape, murder, and theft, are prohibited by law in every country around the globe. Such behaviors are universally abhorrent and are punished accordingly. Civilized, conscientious people obey the rules of their society, but within every society are some individuals with a propensity for rule-breaking. Often these people commit heinous acts with a shocking lack of remorse or compassion, thus violating the cooperative rules of societies. Because these behaviors are statistically abnormal and cause suffering to others, they are often deemed clinically pathological. Antisocial personality (here used interchangeably with the terms *sociopathy* and *psychopathy*) is a diagnosis that captures this phenomenon.

A thorough accounting for the current diagnostic criteria for sociopathy is in order. Note particularly how these symptoms are socially oriented (as the name implies) and consider the implications for a society of cooperators. To be considered a sociopath, a person must exhibit the following:

(1) failure to conform to social norms with respect to lawful behaviors as indicated by repeatedly performing acts that are grounds for arrest;
(2) deceitfulness, as indicated by repeated lying, use of aliases, or conning others for personal profit or pleasure;
(3) impulsivity or failure to plan ahead;
(4) irritability and aggressiveness, as indicated by repeated physical fights or assaults;
(5) reckless disregard for safety of self or others;
(6) consistent irresponsibility, as indicated by repeated failure to sustain consistent work behavior or honor financial obligations;
(7) and a lack of remorse, as indicated by being indifferent to or rationalizing having hurt, mistreated, or stolen from another.[36]

Prisons are filled with people with this disorder, including serial killers, serial rapists, and con artists; unsurprisingly, this disorder is much overrepresented in males. There is an interdisciplinary body of evidence supporting the heritability of antisocial personality disorder,[37] suggesting that sociopathic behaviors may have been selected for, particularly in the evolutionary past of our male ancestors. A hallmark of the disorder is engaging in social deception. In a classic paper on the sociobiology of sociopathy, Linda Mealey described sociopaths as:

> individuals of a certain genotype, physiotype, and personality who are incapable of experiencing the secondary "social emotions" that normally contribute to behavioral motivation and inhibition; they fill the ecological niche described by game theorists as the "cheater strategy."[38]

In other words, those with antisocial personality disorder are the ultimate cheaters. Not only are they prone to aggression but they also tend to perpetrate fraud, pretending to be cooperative under false pretenses, usually as a means to steal from others.

Stealing is not only considered pathological by the standard of the American Psychiatric Association, but also by the Abrahamic god. Through commandment number eight (*though shall not steal*), the Judeo-Christian god made it clear that stealing was a prohibited behavior. There are many other references in the Bible addressing the problem of stealing:

> You shall not steal; you shall not deal falsely. (Lev. 19:11)
>
> If a man steals an ox or a sheep and slaughters it or sells it, he must pay back five head of cattle for the ox and four sheep for the sheep. (Exod. 22:1)
>
> The LORD hates dishonest scales but accurate weights find favor with Him. (Prov. 11:1)
>
> Let him that stole steal no more: but rather let him labour, working with *his* hands the thing which is good, that he may have to give to him that needeth. (Eph. 4:28)

While these rules seem morally sensible, they are not in fact unconditional; if the rest of the Bible is any guide, they were meant only to pertain to the in-group, arguably as rules of reciprocity aimed at maintaining group cohesion and cooperation. As for the out-group—it is fair game. Biblical patriarchs, for instance, were known to steal from outsiders, and such stealing was encouraged by God, even if it meant slaughtering the victims:

> Whenever David attacked an area, he did not leave a man or woman alive, but took sheep and cattle, donkeys and camels, and clothes. Then he returned to Achish. When Achish asked, "Where did you go raiding today?" David would say, "Against the Negev of Judah" or "Against the Negev of Jerahmeel" or "Against the Negev of the Kenites." He did not leave a man or woman alive to be brought to Gath, for he thought, "They might inform on us and say, 'This is what David did.'" And such was his practice as long as he lived in Philistine territory. (1 Sam. 27:9–11)

In the following example, God gives Jericho to Joshua. Joshua steals all the valued possessions of the citizens of Jericho and commits genocide in the process:

> Joshua said to the people, "Shout; for the LORD has given you the city. And the city and that is within it shall be devoted to the LORD for destruction. . . . But all silver and gold, and vessels of bronze and iron, are sacred to the LORD; they shall go into the treasury of the LORD."

. . . Then they utterly destroyed all in the city, both men, and women, young and old, oxen, sheep, and asses, with the edge of the sword. . . . And they burned the city with fire, and all within it; only the silver and gold, and the vessels of bronze and iron, they put into the treasury of the house of the LORD. . . . So the Lord was with Joshua, and his fame spread throughout the land. (Josh. 6:16–27)

The Koran also prohibits stealing in no uncertain terms—"Now as for the man who steals and the woman who steals, cut off the hand of either of them in requital for what they have wrought, as a deterrent ordained by God: for God is almighty, wise" (Koran 5:38)—but it also holds the same double standard. Mohammed funded his very rise to power with wealth he sequestered from raiding desert caravans. He personally took part in twenty-seven raids, which are widely regarded by historians as having been offensive in nature (rather than in self-defense) and undertaken as a means to acquire resources.[39] The spread of Islam would likely not have taken place without Mohammed's raids.

The ethical insularity of in-group logic, whether in religious or secular efforts to steal resources, seems to play out unconsciously, which evolutionary psychologists would regard as a sign of its evolved design. The same is true for killing.

SOCIOPATHIC KILLING

In the summer of 1969, Charles Manson ordered his followers to murder the inhabitants of a Los Angeles suburban home, saying "totally destroy everyone in it, as gruesome as you can." Sharon Tate, a young Hollywood actress, was among the victims. Though Tate was pregnant at the time and desperately pleaded for the life of her unborn child, she was stabbed sixteen times. With the blood of Tate's corpse, Manson's disciples smeared the word "Pig" on the front door of the home.[40] Ted Bundy was another infamous sociopath who raped and killed at least thirty women, twelve of whom he decapitated. Ahmad Suradji was an Indonesian serial killer who murdered forty-two women and young girls. He was known for burying his victim up to their waists before strangling them with a cable. The list goes on, but I will not belabor the point.

If in reading these accounts you get cold chills up your spine or a sense of moral disgust, your moral compass is working properly. Acts so vile and so lacking in remorse, whether committed personally or ordered by a dominant individual, violate a code of morality so completely that in many societies these criminals are deemed undeserving of life—two of these men were executed with much support from the societies in which their atrocities were committed. Clinically, the behaviors of these men are the quintessence of sociopathy.

But this diagnosis, this class of cheaters so universally abhorred by in-group members, may lose its horror when the acts are committed against the outsider, as the outsider often fails to benefit from the established moral codes underlying prohibitions against rape, murder, torture, and infanticide. Again, men don't require religion to dichotomize morality in this manner. But among religious institutions, which worldwide are considered pillars of morality, dichotomous morality is enunciated in the most basic of foundational credos. We saw this earlier in Teehan's observation that "thou shall not kill" really means "thou shall not murder those with whom you have cooperative agreements." Hypocrisy then, it would seem, is another violation of the moral code applying only to in-group cooperators.

I argue that bias toward the in-group, and toward the dominance structures therein, are among the most deeply embedded and most dangerous characteristics of the human race. These traits underlie warfare, oppression, torture, and other cruelties and allow most humans to be incredibly skilled at moral hypocrisy, given the right in-group–out-group manipulation. Of self-righteous bias, David Livingstone Smith observes:

> Self-deception lubricates the psychological machinery of slaughter, providing balm for an aching conscience. By pulling the wool over our own eyes and colluding with our own deception, we can continue to think of ourselves as compassionate, moral and pious people, while endorsing or participating in the wholesale destruction of other human beings.[41]

While in-group bias is clearly a quality to which humans are already prone, religion too often legitimizes and encourages it in action. The Judeo-Christian god, for example, directly orders that compassion for fellow humans be suspended, perhaps as a means to lift the brakes on intraspecies violence (italics mine):

> And thou shalt consume all the people which the LORD thy God shall deliver thee; *thine eye shall have no pity upon them.* (Deut. 7:16)

With compassion removed from the equation, all manner of killing is possible. Genocide, for instance, is not uncommon in the Bible. Through Moses, God helps Joshua decimate twenty kingdoms (as estimated by religious critic Steve Wells)[42] and give the spoils to the Israelites:

> He totally destroyed them, as Moses the servant of the LORD had commanded. . . . The Israelites carried off for themselves all the plunder and livestock of these cities, but all the people they put to the sword until they completely destroyed them, not sparing anyone that breathed. (Josh.11:12, 14)

Here is a snapshot of other acts of genocide committed on the order of God, reminiscent of the crimes of Manson and his disciples (killing men, women, and unborn children), only on a much greater scale:

> Thus saith the LORD of hosts . . . go and smite Amalek, and utterly destroy all that they have, and spare them not; but slay both man and woman, infant and suckling, ox and sheep, camel and ass. (1 Sam. 15:2–3)

Genocide is truly one of the diseases of human existence, if only in its sheer capacity to cause human suffering. And yet scriptures praising genocide continue to be treated as holy texts in churches and cathedrals around the world because the killings are considered triumphs against evil, commanded as they were by a righteous god who is by his nature infallible.

The Crusades, too, were considered righteous campaigns in the era in which they were fought, even recalled centuries later by world leaders such as George W. Bush who called for a "crusade on terror." In reality, the Crusades were a long series of battles with mindboggling destruction—much of it by Christians. As one example, on July 15, 1099, about twelve thousand crusaders descended upon Jerusalem, breeched the city's walls, and tore the city apart. The eyewitness account of priest Raymond of Aguilers should remind us of the strong in-group bias pervading human morality:

> Wonderful things were to be seen. Numbers of the Saracens were beheaded ... others were shot with arrows, or forced to jump from the towers; others were tortured for several days and then burned in flames. In the streets were seen piles of heads and hands and feet. One rode about everywhere amid the corpses of men and horses.[43]

Other documents wrote of women being raped and stabbed, suckling babies being wrenched from their mother's breasts and slammed against posts or thrown over the walls. As many as seventy thousand Muslims were killed, and the Jews who remained were thrown into a synagogue and torched alive.[43]

Islam is just as guilty of reverse logic on killing when it concerns the out-group, and the Koran seems ripe with passages that recount unspeakable acts in the name of religion. But first, as in like passages of the Judeo-Christian Bible, compassion must be set aside:

> Be not weary and faint-hearted, crying for peace, when ye should be uppermost [have the upper hand] for Allah is with you. (Koran 47:35)

With this out of the way, all violence becomes possible:

> I will cast terror into the hearts of those who disbelieve. Therefore strike off their heads and strike off every fingertip of them. (Koran 8:12)

> The punishment of those who wage war against Allah and His messenger and strive to make mischief in the land is only this, that they should be murdered or crucified or their hands and their feet should be cut off on opposite sides or they should be imprisoned; this shall be as a disgrace for them in this world, and in the hereafter they shall have a grievous chastisement. (Koran 5:33)

If this kind of religion-inspired brutality simply reflected the sensibilities of a bygone era, a set of moral perspectives that we have matured out of since the biblical age, that would be one thing, but acts of remorseless killing or torture or mutilation in the name of religion are not confined only to the Abrahamic faiths, or to the ages of Abraham, Jesus, or Mohammed. In 2002, for instance, Hindu mobs in Gujarat, India, descended upon the Muslim out-group that lived there:

> Mothers were skewered on swords as their children watched. Young women were stripped and raped in broad daylight, then doused with kerosene and set on fire. A pregnant woman's belly was slit open, her fetus raised skyward on the tip of a sword and then tossed onto one of the fires that blazed across the city.[44]

The mob rampaged through the Muslim neighborhood, stealing, raping, and burning 124 Muslims alive. These chilling acts bear the marks of primate territorialism, in-group loyalty, alliance with the alpha, and primate displays of dominance. The Hindu mob's atrocities were reportedly linked to a previous attack on a train car filled with activists from Vishva Hindu Parishad (the World Hindu Council). As the story goes, an ancient sixteenth-century Mosque was demolished to make room for a Hindu temple dedicated to the god Ram (a male supreme being and warlord). Stirred by their recent campaigning for construction of the temple, the council members rushed from the train, attempted to force a Muslim vendor to say "Hail Ram," pulled his beard (we will discuss the evolutionary significance of beards in the next chapter), and beat him when he refused. Muslim mobs retaliated, attacking the train with stones, setting it on fire, and killing fifty-nine people. Most of the victims on the train were reportedly women and children. People operating under the rules of in-group loyalty and commitment are generally shocked by such acts of femicide and infanticide, though men feel much less obligation to see such actions as immoral when they are committed against the enemy.

For the best example of the double standard in religious killing, we must look to the Judeo-Christian god himself. The god of the Bible—while killing for reasons defensible by the standards of in-group morality—murders in the manner of the most remorseless sociopathic killer. In his book *Drunk With Blood: God's Killings in the Bible*, Steve Wells highlights the caprice of God's killings: "God buries the opposing party alive along with their families" (Num. 16:34); "God burns 250 alive for burning incense" (Num. 16:35); "God killed 14,700 for complaining about God's killings" (Num. 16:49).[45] One could—and Wells does—populate many pages with similar examples. However, what is perhaps most revealing in Wells's work is his simple summation of all the killings numerated throughout the Bible, including the number attributed to God versus those attributed to Satan. God, it turns out, is directly described as killing two and a half million people—2,475,636 to be precise. Satan is listed as killing ten. When Wells includes estimates accounting for the probable size of towns and communities God is purported to have decimated, God killed over 24 million people (24,643,205)—and many were his own followers—while Satan killed sixty.[46] Needless to say, as God's enemy, the devil is seen as vile, pernicious, and dangerous—the very epitome of evil. Yet the Christian god is seen as a righteous god, and his killings are censored, rationalized away, ignored, or simply disappear behind a wall of denial. But the most important point is that God didn't commit any of these killings. When not a result of mindless forces of nature (such as earthquakes, floods, disease), the killings were committed by the hands of men, as they always are in religious warfare.

The stated reason for religious (and secular) warfare often involves some sort of higher ideal. Such ideals give the religious moral justification for the acts they commit in the fields of battle. However, the engines of religious war, like other kinds of organized violence, have been forged from ancient patterns of primate alliance-making and shaped by the rules of in-group cooperation. God is given to play a central role, evoking the dominant males of our evolutionary history in a manner that is deeply intuitive to our evolved psychology. Nonhuman primates are obsessed with dominants and spend a disproportionate amount of time and energy trying to make alliances with them. Human primates are little different, and with God represented as the most powerful male in the universe, the religious experience an especially powerful pull to form alliances with him. To do so they make costly signals of cooperation, often by attacking his would-be enemies. They also make alliance by defending God's word, sometimes through extreme measures such as killing. At other times, alliance is made by offering sacrifice to God. The worst of these offerings come in the form of "food" from human corpses and blood. In all, following the patterns of primate alliance-making with God has wrought terrible human suffering.

It is clear how coming together under the aegis of deeply intuitive religious ideologies could have benefitted our ancestors, particularly those living in the brutal days during which the Abrahamic religions arose, when the need to rally under dominant males was a matter of life and death. However, like many aspects of our evolved psychology, becoming aware of our unconscious predispositions enables us to choose to abandon those which amplify human suffering. Once the biological mechanisms underlying moral hypocrisy are recognized, it may make it easier to disallow sociopathic behavior, even when it occurs outside the boundaries of our immediate circle. This should also be true for religious circles and should render questionable any "higher" ideals fomenting religious violence. Loyalty to a particular dominant male god should perhaps bear the most scrutiny, as such loyalty is often the ultimate mark of religious in-group identity and the progenitor of the most merciless atrocities against the out-group.

Chapter 6

WHAT IT MEANS TO KNEEL

> At first sight it is surprising that religion has been so successful, but its extreme potency is simply a measure of the strength of our fundamental biological tendency, inherited directly from our monkey and ape ancestors, to submit ourselves to an all-powerful, dominant member of the group.
> —Desmond Morris, *The Naked Ape*

SIZE AND DOMINATION: WHAT IT MEANS TO BE BIG

Our primate ancestors passed on to us certain social protocols, and we have passed them on to God. In primate cliques, subordinates will often shrink down before the dominant male, thus accentuating his largeness and superiority. The god of the Abrahamic religions, in company with gods of other traditions, is often portrayed as a large male who requires that subordinates lower themselves before him. However, requirements like these violate central tenets of the Abrahamic god concept such as omnipotence (he should have no truly viable competitors), incorporeality (size should only matter to a being on the physical plane), and immortality (God does not need to reproduce, therefore subordinating other males holds no reproductive value). To understand this projection, we begin with the role of size in primate hierarchies.

Sex differences in size are attributable to sexual selection acting by way of male mate competition.[1] Larger males are generally better at winning mate competitions, and they pass on their genes for larger size to subsequent generations of males. Humans evolved in social hierarchies where rank—particularly among males—was often contested with violence. Because of the high costs associated with physical confrontation, the ability to perceive the dominance rank of potential rivals is seen as having been an important selective pressure (i.e., *not* accurately detecting high rank can be deadly) that shaped the human brain.[2] Indeed certain brain structures appear to be devoted to processing rank information.[3] As such, humans process rank information unconsciously and with incredible speed.[4] Understanding the relationship between size and the power to cause physical harm is a critical part of recognizing dominance.

Larger size often equates to dominance in many other species[5] and plays an important role in the rank structures of men. As cognitive scientist and linguist Steven Pinker has pointed out, the big-men who ruled over hunter-gatherer societies were often literally *big men*.[6] Even though size may have less significance in competitions among contemporary men in the industrialized societies of the world (where resources aren't necessarily won through physical means), size still bears psychological significance. One example comes from the workplace. Research has found that height plays an important role in labor markets; it is correlated with dominance, status, and higher earnings and can even affect whether one holds a blue collar or a white collar position. For instance, statistically sales managers are taller than salesmen, and bishops, taller than preachers.[7] Height in men has also been linked with physical strength,[8] fighting ability,[9] social status,[10] and even reproductive success.[11] It is important not to confuse correlation with causation here, and there are certainly other variables influencing dominance among men, including intelligence, health, and charisma, to name a few. However, across the animal kingdom it usually pays to be larger. Most animals understand this and use size to make inferences about things such as power, dominance, and threat potential. Some will even feign greater size in the hopes of submitting a rival.

Big Heads, Big Hats

To appear larger, many animals possess adaptations that exaggerate their head size. We see this ancient strategy across species. For instance, frill-necked lizards have skin flaps on their heads for flaring open in the face of rival males (or predators). Male lions have fantastic manes for similar purposes. Elephants threaten by spreading their huge ears, making their large heads appear even larger.

Primate males also possess these adaptations. The male orangutan has a zygomatic bone that flares out from either side of its head around the upper jaws, large fatty cheek pads, and a long orange beard. These appurtenances are designed to convey dominance and make a mature male's head appear enormous compared to a female's or an adolescent male's. As with the orang, facial hair is used for intimidation displays in other primates. The howler monkey has a long hanging beard, baboons have great tufts around the sides of their heads, and marmosets have large ear tufts.

The relevance of such displays has not bypassed modern-day men. Paleobiologist R. Dale Guthrie has observed that beards in men appear to extend the edge of the chin and make the head seem bigger, which may serve similar purposes as in other primates.[12] Indeed, research finds that subjects rate images of men with beards as more aggressive[13] and as having higher social status[14] than images of clean-shaven men. Like other sexually dimorphic traits, beard growth in men is stimulated by testosterone, a hormone associated with aggression.[15] A look at the symbolic role beards play in dominance among men (particularly high-ranking or fighting men) further supports Guthrie's hypothesis. For instance: motorcycle gang members often sport beards to look threatening; knights of the Middle Ages sported beards, which were symbols of virility and honor—holding someone else's beard was considered an offense righted by a duel; and kings across history wore beards as symbols of dignity and honor, virtues that convey high status.

Because power and dominance are pervasive themes across religious traditions, it should come as no surprise that a primitive adaptation designed to intimidate rivals would have such a common place in religious culture, nor that it retains its primeval meaning. Saint Augustine, for instance, wrote that "the beard signifies the courageous."[16] Saint Clement of Alexandria said the beard lent the face "dignity and paternal terror"[17] and wrote that God "adorned man, like the lions, with a beard, and endowed him, as an attribute of manhood, with shaggy beasts—a sign of strength and rule."[18] Given the symbolic power of beards, religious men have sported them widely. Virtually all sects of Islam encourage men to wear beards, and religious prescriptions for long beards are referenced in Hinduism, Rastafarianism, and Sikhism. Judaism is no exception—the Old Testament, for instance, orders that,

"You shall not round off the side-growth of your heads nor harm the edges of your beard" (Lev. 19:27). Some Christian clergy, including the Orthodox, are required to wear long beards, as are the Amish. More directly linking beards to dominance competition, renegade groups of Amish men in Ohio have taken to assaulting their brethren and chopping off their long beards.[19] Further exemplifying the connection to dominance, pharaohs of ancient Egypt wore *postiches*, exaggeratedly long false metal beards which were symbols of divine status and power. Finally, in American popular culture, God is sometimes referred to as the "the bearded old man," which connotes a different level of dominance altogether from, say, a *smooth-faced adolescent*.

While most primates exaggerate head size using the parts nature supplied them (e.g., beards, fatty cheek pads), humans also create artificial head displays to signify dominance. Plains Indian warriors wore horned buffalo headpieces specifically to intimidate rivals in battle. Armed soldiers of Britain's Queen's Guard wear big hats—not made of felt or straw, but of *black bearskin*. Here we have armed men guarding a high-status woman wearing the skin of a fearsome predator on their heads. Kings across the ages wore spiked crowns of gold adorned with precious stones which convey status and wealth, a common proxy for status in humans. The despotic Kafa chiefs of Ethiopia wore soaring three-foot-tall conical headpieces adorned with the most blatant example of mate competition among men—three golden phalluses.[20]

On the flip side, removing one's hat is regarded as a sign of deference, submission, and respect. Across history, hats have been removed when addressing people of nobility or high status. In the twelfth century, after the Mongol hordes hacked Russia to pieces, Russian princes demonstrated submission by removing their hats, filling them with grain, and feeding the Mongol horses from them. Only after this show of humiliation did the Mongols halt their rapacious assaults on the Russian people. In Christianity it is considered disrespectful for men to wear hats in church; accordingly, before entering the house of God, men remove them. Women do not generally have the same obligation, as we would expect if the gesture were rooted in specifically male mate competition.

As with beards, religious men capitalize on the impact of big hats, which resonate with deep, evolved intuitions about status. The only men allowed to wear big hats in Christian churches are clergy, men high in the religious hierarchy. Buddhist lamas wear large headdresses—one type looks like a giant version of the helmets worn by the ancient Roman army. The Dalai Lama wears one. Even in spiritual philosophies founded on humility and egalitarianism there is rank, and the high-ranking cannot seem to escape the pull for large head displays. Kohen Gadol, orders of high priests of classical Judaism, wore "mitznefet"—large mushroom-shaped turbans topped with gold crowns. Lower-order priests wore much smaller, plainer cone-shaped hats that were clearly intended to contrast with those of their superiors. Highlighting how religious customs are rooted in male mate competition, the Bible describes how the big-hatted Kohen Gadol enjoyed the exclusive privilege of marrying virgins (Lev 21:13).

The pope of the Roman Catholic Church wears a giant, pointed hat called a mitre. In Church hierarchy, the mitre is directly tied to status, restricted to the pope, bishops, and other high-level clergy. Some of the popes' mitres are made of gold and, like the crowns of kings, are dotted with precious jewels. Popes also wear the *papal tiara*, or the *triple crown*, another coronet of gold and sparkling gems. The stated meaning of the crown is significant—it is meant to indicate power and rulership and shows that as God's representative, the pope's power transcends civil authorities. Here is the traditional pronouncement of papal coronation:

> Receive the tiara adorned with three crowns and know that thou art the father of princes and kings, ruler of the world, vicar of our savior Jesus Christ on earth, to whom is honor and glory in the ages of ages.[21]

While gods aren't typically depicted wearing hats, they are frequently shown in religious iconography with large circles of light surrounding their heads—known as halos, or *glory*. Not only do halos appear to extend the head size, but they have been linked to the prowess of male warriors in mate competition. In the *Iliad*, for example, Homer describes warriors in mortal combat having supernatural light surrounding their heads (recall that these males were, perhaps symbolically, fighting over the possession of a female, Helen of Troy). In Greek iconography, heroes (such as Perseus while slaying Medusa) are sometimes depicted with halos. The gods of Sumerian religions are described as having halos, as are Sumerian kings and legendary Sumerian male heroes. Angels, God's armed and fearsome warriors, are synonymous with halos. Not surprisingly, Greek and Roman gods, the Buddha, and Ra the Egyptian sun god are all depicted with halos, as are God's envoys Abraham, Muhammad, and Jesus Christ.

Meanwhile, big hats are also important symbols in spiritual warfare. In Revelations, Jesus rides in on a white horse and makes war on Satan, the False Prophet, and all the kings and generals of the earth. His robe is bloodied (some say from his vanquished enemies), and on his head he wears a stack of crowns: "His eyes are like blazing fire, and on his head are many crowns" (Rev. 19:12). After slaughtering his enemies, Christ becomes the absolute ruler of the earth (Rev. 19:11–17). Also in Revelations, when angels submit kings to the eternal power of God, the kings remove their big hats (italics mine): "They *cast their crowns before the throne*, saying, 'Worthy are you, our Lord and God, to receive glory and honor and power, for you created all things, and by your will they existed and were created'" (Rev. 4:9–11). It is safe to say that the kings of Revelations did not require, as a condition of their surrender, that God also remove *his* halo, or Christ, his stack of crowns.

Posturing

To avoid costly physical confrontations, many species demonstrate their rank status through posturing—rising up and projecting greater size to convey dominance, or shrinking down and projecting smaller size to convey submission. Making oneself appear smaller may be a gesture designed to convey infantile qualities—most animals are programmed not to attack their young.[22] Chimpanzees convey dominance by bristling their hair and standing upright like men. Likewise, when they wish to show submission they shrink down, faces planted to the ground. Frans de Waal offers a rich description based on his observations at Burgers' Zoo in Arnhem:

> Strictly speaking a "greeting" is no more than a sequence of short, panting grunts known as pant-grunting. While he utters such sounds the subordinate assumes a position whereby he looks up at the individual he is greeting. In most cases he makes a series of deep bows that are repeated so quickly one after the other that this action is known as

bobbing. Sometimes "greeters" bring objects with them (a leaf, a stick), stretch out a hand to their superior or kiss his feet, neck, or chest. The dominant individual reacts to this greeting by stretching himself up to a greater height and making his hair stand on end. The result is a marked contrast between the two apes, even if they are in reality the same size. The one almost grovels in the dust, the other regally receives the greeting. Among adult males this giant/dwarf relationship can be accentuated further still by histrionics such as the dominant ape stepping or leaping over the "greeter." ... At the same time the submissive ape ducks and puts his arms up to protect his head. This kind of stuntwork is less common in relation to female greeters. The female usually presents her backside to the dominant ape to be inspected and sniffed.[23]

From de Waal's description we can draw many religious parallels. British zoologist Desmond Morris offers his own insights about religious submission, which have a striking concordance with the behaviors observed by de Waal (although made decades before de Waals's observations):

> We are forced to the conclusion that, in a behavioral sense, religious activities consist of the coming together of large groups of people to perform repeated and prolonged submissive displays to appease a dominant individual ... the submissive responses to it may consist of closing the eyes, lowering the head, clasping the hands together in a begging gesture, kneeling, kissing the ground, or extreme prostration, with the frequent accompaniment of wailing or chanting vocalizations. If these submissive actions are successful, the dominant individual is appeased. Because its powers are so great, the appeasement ceremonies have to be performed at regular and frequent intervals, to prevent its anger from rising up again.[24]

Not only are such submissive behaviors built into religious ritual, particularly when approaching God or his proxies, but the contrast of a higher being and lower supplicants is often built into the physical architecture of religious and political power. The thrones of kings, the pulpits of presidents, and the altars of priests are all raised high above the masses. Conceivably, religious men could sermonize in a pit, something like the floor of the Colosseum—everyone would still be able to see—but altars, like the soaring cathedrals across the globe in which they are housed, often tower in majesty, which is crucial for their psychological impact. The throne of God, at the foot of which kings relinquish their crowns, is a towering emblem: "I saw the Lord sitting upon a throne, high and lifted up" (Isa. 6:1).

The values we place on the vertical dimension is conspicuously reflected in our language—"high priests," "high and mighty," "higher order," "high heaven," as opposed to "lowly heathens," "lower rank," "to occupy a low station in life." In reality, space is a valueless and relative dimension, apart from our projections as animals that ascribe hierarchical status to stature. Yet God is high up in the heavens, whereas Lucifer is down in hell; as primates, we couldn't rightly have Lucifer sitting higher than God. When Lucifer rises up in an attempt to show dominance, God forces him down again, following the pattern established by the earth's competitive animals. Here is how it is described in the Bible (italics mine). When Lucifer proclaims:

> I will *ascend* into heaven,
> I will exalt my throne *above* the stars of God;
> I will also sit on the *mount* of the congregation
> On the farthest sides of the *north*;
> I will *ascend above the heights* of the clouds,
> I will be like the *Most High*.

God's responds:

> Yet you shall be brought *down* to Sheol,
> To the *lowest depths* of the *Pit*. (Isa. 14:13–15)

It seems illogical that God should need to be large to exert his dominance given that he is capable of creating the physical universe by simply speaking it into being—concerns about size reflect the realm of men, bound to the rules of the physical world. Nevertheless, the Bible illustrates his "highness," a term that in addition to meaning *majesty* also alludes to largeness and height, features that are meaningful to primate rank and power. Here God towers over the earth and uses the whole planet on which to rest his feet:

That men know that you, whose name alone is JEHOVAH, are *most high* over all the earth. (Ps. 83:18)

However, the *most High* dwells not in temples made of hands: as said the prophet, "Heaven is my throne and the earth my footstool..." (Acts 7:48)

Exaggerating size differences also has a special place in Islam. Submission to Allah is demonstrated through prostration whereby the pious place their nose and forehead onto the ground and recite (italics mine), "Glory to my Lord, *the Most High,* the Most Praiseworthy" in succession. The centrality of submission among Muslims is understood with the very translation of the world *Islam,* which literally means *submission.*

As concerns prostration, bowing, and the like, there are too many references in the Bible to list comprehensively here, but the following is representative:

> Come, let us bow down in worship,
> let us kneel before the LORD our Maker;
> for he is our God. (Ps.95:6–7)

Kneeling, as noted in this verse, offers yet another means of lowering oneself. Bowing and kneeling are behaviors practically synonymous with showing deference to dominant humans. Since ancient times, kneeled stances have been taken in the presence of dignitaries, kings, and others of high status and remain common in societies with large power or class differentials. Kneeling is also customary across many religions. The Catholic Church, for one, abounds with kneeling rituals. In church, when one crosses the altar's corridor, one is to kneel. In the presence of a bishop or pope, one is to kneel. Naturally, when one is addressing God in prayer, one is to kneel. Ritualistic submission has even etched its way into Christian interior design, where behind every pew is a narrow bench on which the devoted are expected to submit themselves by kneeling.

Eye Contact

Living in a dominance hierarchy requires that members have the cognitive ability to understand the status of those around them and to anticipate their behavior and emotional state.[25] Primates are particularly adept at these cognitive tasks and their brains have evolved to infer this information from eye contact. Studies of nonhuman primates have found neurons specifically designed to detect eyes gazing toward the viewer.[26] Humans also possess these adaptations, and research suggests they are related to emotions central to navigating rank status. For instance, research in humans has found that the amygdala, the fear and anger center of the brain, is a critical region for monitoring gaze direction.[27] Gaze direction is complex and can be related to things like affiliation, securing support, or outside danger. However, in many species, including human and nonhuman primates, a direct stare is often meant to indicate a threat, whereas averting the eyes is meant to indicate submission.[28]

Humans exhibit rules for eye contact that vary by rank status, perhaps most notably in highly stratified societies. In the presence of the despotic chiefs of the south Pacific island of Tikopia, for example, the eyes of subordinates were to be averted. As cited by anthropologist Laura Betzig, ethnographer Raymond Firth notes what would happen if a subordinate made eye contact with a dominant Tikopian chief: "If a chief catches sight of an upturned face as he strides onto the marae, he calls out to the offender 'Who is the person who looks onto the *fono* of the gods?'" The offender is overcome with disgrace and shame, and then commences the "histrionics," to borrow from de Waal—he paddles out to sea in his canoe (a symbolic act of suicide), then returns to his own gardens, packs the canoe with food (offerings), and returns to the chief where he begins kissing the chief's feet and making submissive hooting noises. Then, "wailing in his humility, he crawls to the chief over floor mats, presses his nose to the chief's feet and knee and follows this by the chanting of a dirge."[29] In the Fur society of Sudan it was also forbidden to look despotic leaders in the face. In fact, it was customary for the king to come out with his face half veiled. In Fur culture, even the king's topmost men were not allowed to gaze upon him.[30]

Averting the eyes is a common, almost unconscious ritual in religious practice. When communicating with God through prayer, eyes are lowered to the ground, heads are kept down. This simple act ties strongly to the rules governing the interactions of male primates. To look a powerful male in the eye when addressing him can have dangerous implications. When Moses encounters God in the burning bush, it is said that "quaking with fear Moses did not dare gaze" (Acts 7:32). Similarly, God himself forbids his followers from looking directly on him, on the threat of death:

> The LORD descended to the top of Mount Sinai and called Moses to the top of the mountain. So Moses went up and the LORD said to him, "Go down and warn the people so they do not force their way through to see the LORD and many of them perish. Even the priests, who approach the LORD, must consecrate themselves, or the LORD will break out against them." (Exod. 19:20)

God repeats the threat again, even more directly: "you may not look directly at my face, for no one may see me and live" (Exod. 33:20). Similarly, in Luke, a tax collector shows submission to God by averting his eyes, hitting himself, and begging for mercy: "He would not even look up to heaven, but beat his breast and said, 'God, have mercy on me, a sinner'" (Luke 18:13).

With religion's emphasis on dominance rank and power, direct eye contact is seen as a challenge, and averting it, a show of submission. Bowing the head is another means to avert the eyes and make oneself smaller at the same time. Across the world today, this gesture is a common form of addressing aristocracy or other high-ranking members of society. It is also common in religious ritual, particularly while addressing high-ranking religious men, or God.

Hand and Foot Kissing

Lip smacking—the nonhuman primate equivalent of kissing—is a common appeasement display in monkeys and apes and may be intended to emulate infant suckling noises. It is understandable how sounds associated with nurturing might appease aggression. As with other infantile behaviors, this gesture may also communicate, "Like an infant, I pose no threat."

Like eye contact, the human kiss has numerous meanings and functions, and humans similarly use kissing as a means to show submission. Throughout the *Godfather* film series, kissing the don's hand was used to signify acknowledgment of his rank status. Whether real dons get their hands kissed, I don't know, but the gesture was perfectly intuitive to viewers. Hand kissing is a customary way of showing submission to a king, typically while bowing one's head or kneeling down. Often the custom was—and is, in the remaining monarchies of the world—to kiss the king's signet ring. This custom remains strong in the Catholic Church, a monarchic hierarchy in which the pious kneel before the pope and kiss his ring. Other Church customs would suggest that this gesture is rooted in ancient primate displays intended to connote infanthood—for instance, the pope is referred to as father, and his flock are considered his children. In many Christian Orthodox churches it is still custom for laity to bow their heads profoundly and say, "Father bless" (to a priest) or "Master bless" (to a cardinal) while outstretching their right hand. The clergyman then performs the sign of the cross and grabs the supplicant's hand, allowing the opportunity for his own to be kissed. Likewise, in letters to clergy one is to open with "Father Bless" and close with "Kissing your right hand." With their remarkable skills at abstract thinking, humans have carried ancient submission displays forward to written language.

Foot kissing, a behavior observed widely in primatology, is another way submissive monkeys and apes demonstrate acquiescence to dominant members of their societies.[31] This behavior carries forward to human societies that are highly rank structured, such as monarchies. For instance, kissing the king's feet has always been synonymous with supplicant behavior—for example, showing him extreme deference, begging for his mercy, or even recognizing that he represents God.

Christ—who is sometimes referred to as *Christ the King*—is also greeted with foot kissing, as are his proxies. At the Basilica in Rome stands a large bronze statue of Saint Paul, built in the fifth century. Though this statue has stood stalwart now for fifteen centuries, its feet have been worn thin by the lips of pilgrims. There was even a custom in the Catholic Church of kissing the feet of the pope. The custom was actually made into law by Pope Gregory VII in his *Dictatus Papae* (Dictates of the Pope). In this document, the connection between foot kissing and rank are made perfectly transparent, lest we confuse the gesture as something more affectionate. Two of the dictates were:

9. That of the pope alone all princes shall kiss the feet.
12. That it may be permitted to him to depose emperors.

These statements reflect an ageless hierarchical arrangement in which the Church sought to wield power even above kings (princes, emperors, etc.). To wit, Christ is also known as the *King of Kings*. Further, with God's backing, even the lowly pious may be treated as kings of kings, in the manner of dominants and their affiliates. The prophet Isaiah describes below how kings submitting to God demonstrate their submission by kissing (or licking, as it were) the feet of God's followers:

> Kings . . . shall bow down to thee with their face toward the earth, and lick up the dust of thy feet. (Isa. 49:23)

Lastly, foot-kissing occurs in one of the best-known stories of the Bible in which Jesus is invited to dinner by Simon, one of the Pharisees. A sinful servant woman begins to attend to him:

> As she stood behind him at his feet weeping, she began to wet his feet with her tears. Then she wiped them with her hair, kissed them and poured perfume on them. (Luke 7:38)

When Jesus's host Simon—perhaps more cognizant of the power implications of such behaviors—refuses to make the same gesture, Jesus becomes indignant.

> Then he turned toward the woman and said to Simon, "Do you see this woman? I came into your house. You did not give me any water for my feet, but she wet my feet with her tears and wiped them with her hair. You did not give me a kiss, but this woman, from the time I entered, has not stopped kissing my feet. . . . Therefore, I tell you, her many sins have been forgiven. (Luke 7:44–47)

It is worth reiterating that such displays are fundamentally submissive in nature, intended to secure the favor of a more powerful being. As if to hammer the point home, Jesus reminds the woman, in front of the Pharisee, of the eternal punishment she averted with her submission display: "Your faith has saved you; go in peace" (Luke 7:48).

From the rank displays of the biblical dinner party, we move to those occurring in the fields of combat, where male-on-male competition is at its most brutal. We start with Odysseus, who cast aside his spear and shield in surrender, removed his hat, and kissed the knees of the enemy king.[32] Similarly, a Hadith (text recounting the deeds and sayings of Muhammad) tells how Muhammad's cousin who fought against him in battle kisses Muhammad's feet to show acquiescence:

> "O Prophet of Allah, this is your cousin, Abu Sufyaan, please be happy with him." Prophet accepted the intercession of Abbas and said, "I am pleased with him. May Allah forgive all the enmity he showed against us." Thereafter, Prophet turned to Abbas and said, "Verily he is our brother." Abbas said "I kissed his [Prophet's] blessed foot while he was seated on his camel."[33]

Another example of foot kissing in Islam is recounted when the Prophet becomes annoyed with his follower Umar, who kisses his feet to avert his wrath (while repeatedly begging his forgiveness):

> Then Umar stood up and kissed the blessed feet of the Prophet and said: O Messenger of Allah, we are pleased with Allah being Sustainer, you being the Prophet, Islam being the Deen and the Holy Quran being the Guide. Forgive us. Allah would further be pleased with you. So Umar kept on saying it till the Prophet became pleased.[34]

SUBMISSION BY IDEOLOGICAL SURRENDER

Thus far we have seen how apes, men, and gods share nonverbal forms of communicating dominance and submission. With their genius for abstract thought, humans also devise ideologies to reinforce hierarchy, such as those expressed in religious dogma. However sublime such ideologies may appear on the surface, they serve the primitive dominance strivings of men. As such, men may enforce their ideologies with the credible threat of violence.

The result is that humans also show submission by unquestioningly adhering to ideology and show dominance by creating or enforcing ideology. For the largest religious institutions, ideological control adds to an already immense base of power, allowing them to consolidate armies, territory, and economic might spanning nations. Men at the helm of this profusion of power encode ideological control into law, in effect legislating mind control over the masses.

One example of ideological domination is the "divine right of kings," which attests that the king can do no wrong because his earthly power is granted by God. Thus any challenge to the king's policy, his political ideology, or even his behavior is considered sacrilege, sanctioned as it is by the most dominant male in the universe. Proponents of the divine right of kings have often cited scripture as justification, for example: "By me kings reign, and princes decree justice. By me princes rule, and nobles, even all the judges of the earth" (Prov. 8:15–16); and "Let every soul be subject unto the higher powers. For there is no power but of God: the powers that be are ordained of God. Whosoever therefore resisteth the power, resisteth the ordinance of God: and they that resist shall receive to themselves damnation" (Rom. 13:1–2).

Similarly, Christianity has a long history of codified ideological control, with one example among Catholics being the dogma of *papal infallibility*. This dogma was promulgated by the Catholic Church at the First Vatican Council of 1870 and states that any dogmatic teaching that the pope conceives of is infallible. Such teachings are considered to be imparted directly from God and are therefore uncontestable.[35] This is the closest thing to godlike power a human being can have. As recounted above, the decrees of popes have commanded that kings drop to the floor and kiss the papal feet.

The notion of heresy, related to infallibility, has been another long-standing instrument used to maintain religious hierarchy. Heresy can be defined as an opposing belief or position, or a challenge to dogma—where dogma is an established religious doctrine not to be disputed or diverged from. In effect, heresy reflects the proscription against disputing papal infallibility or the divine right of kings. In the grand spectacle of religious history, efforts to forbid heresy, with its mutinous challenge to the dominance of kings and religious leaders, were often coded into law and enforced by execution.

The first Christian to be convicted of heresy was Priscillian, a bishop of Ávila who was executed in 385 by Roman officials for heresy (or, as the civil charges read, for *practicing magic*); religious dogma of the in-group has always been the true religion, whereas the out-group practices "magic." Priscillian was condemned for practices such as allowing women to join with men during prayer and fasting on the Sabbath.[36] Fifteen hundred years later, in 1876, Cayetano Ripoll suffered the last known execution for heresy conducted by the Roman Catholic Church. Ripoll was a Spanish school teacher whose mistake was to teach his students about deism—a religious philosophy that typically rejects reliance on religious authorities and *revealed religion*, the notion that religious dictates are divinely revealed to men like popes and kings. Deism was in many ways a direct threat to the idea of vicarship and Church dominance over the subordinated masses.[37] Beginning in the twelfth century, the Inquisition, or *Inquisitio Haereticae Pravitatis* (inquiry on heretical perversity), was an organized mass effort aimed at combating heresy conducted by the judicial authorities of the Catholic Church. During this centuries-long marathon of intolerance, the Church murdered, mutilated, and tortured hundreds of thousands of people suspected of heresy.

Catholics weren't the only ones to persecute and kill on the basis of heresy. Protestants were also known for executing heretics, most of whom were, naturally, Catholics. Similarly, Orthodox Jews have historically regarded those who stray from the Jewish principles of faith as heretical. Accusations of heresy have also been lodged between rival Islamic sects, most famously between members of the Sunni and Shi'a traditions who regard each other's beliefs as heretical and have been willing to prove it by blowing each other to pieces.

Another means of enforcing mental submission is by prohibiting the reading of religious texts. There is a long history of the Catholic Church prohibiting the masses from reading the Bible. For instance, Pope Innocent III wrote in 1199:

> The mysteries of the faith are not to be explained rashly to anyone. Usually in fact, they cannot be understood by everyone but only by those who are qualified to understand them with informed intelligence. The depth of the divine Scriptures is such that not only the

illiterate and uninitiated have difficulty understanding them, but also the educated and the gifted.[38]

Note the strong reluctance to relinquish power at the heart of the pope's words, which have the effect of forcing the pious to rely on papal interpretation. The practice of prohibiting God's lambs from reading the word of the Lord carried over to the Spanish conquest of the Americas, where indigenous people were not allowed to read scripture. This allowed the Church to practice selective proselytization, thus more easily submitting the natives to ideological control and to kneeling before high (Christian) altars. Masses were often conducted in Latin, an arcane language even for Spanish colonists, which cemented reliance on the priests and furthered their necessary mystique as the translators of God's directives. Thus rank was used to control access to scripture, ultimately to preserve status.

Not only has knowledge, and the questioning it might lead to, been punished, but its spread has often been cannily averted. Book burning has been one reliable means for this; once a popular pastime during medieval Christianity, it has been practiced with great enthusiasm by the modern-day Taliban in Afghanistan. The Koran, one of the few books that incite people to murder when it is burned, provides an extensive recitation of admonitions against freethinking. The most draconian wrath is reserved for unbelievers, those who would question God's word (as written by powerful men). Below is but a minute sampling:

> But the infidels who die unbelievers shall incur the curse of God, the angels, and all men. Under it they shall remain forever; their punishment shall not be lightened, nor shall they be reprieved. (Koran 2:162)
>
> The unbelievers are like beasts [note the infrahumanization] which, call out to them as one may, can hear nothing but a shout and a cry. Deaf, dumb, and blind, they understand nothing. (Koran 2:171)
>
> How steadfastly they seek the Fire! That is because God has revealed the Book with truth; those that disagree are in extreme schism. (Koran 2:175-176)

> As for unbelievers, neither their riches nor their children will in the least save them from God's judgment. They shall become fuel for the fire. (Koran 3:10)

> Believers, if you yield to the infidels they will drag you back to unbelief and you will return headlong to perdition. . . . We will put terror into the hearts of the unbelievers. . . . The Fire shall be their home. (Koran 3:149–51)

Built into the canon are punishments for freethinking that would make virtually anyone hesitant to hold, much less voice, a dissenting opinion. This is even truer for those who believe in divine punishment. The same can be seen in the New and Old Testaments. While Jesus was a man who preached compassion, love, charity, and other humanistic principles, he was also a man who understood the power of fear to motivate men. As such he was not above making threats to dissuade ideological challenge:

> He who rejects me and does not receive my sayings has a judge; the word I have spoken will be his judge on the last day. (John 12:48)

> If a man does not abide in me, he is cast forth as a branch and withers; and the branches are gathered, thrown into the fire and burned. (John 15:6)

Moreover, Christ's ideological power as a dominant male extended to his representatives. Per Luke, if Christ's disciples—who were charged with spreading his word—were not accepted by the communities in which they proselytized, there would be consequences: "It shall be more tolerable on that day for Sodom than for that town" (Luke 10:12); recall that God killed every person in Sodom by raining down burning sulfur upon the city (Gen. 19:24–25).

The wrath of God's armies was reserved for sinners, or those who failed to abide by the moral precepts that he devised: "The Son of man will send his angels, and they will gather out of his kingdom all causes of sin and all evildoers, and throw them into the furnace of fire; there men will weep and gnash their teeth," whereas the righteous "will shine like the sun in the kingdom of their Father" (Matt.13:40–43). Now, suffice it to say that Christ never committed despotic acts such as this during his lifetime—burning heretics alive was something taken up centuries later by his followers, particularly during the Inquisition. However, Christ (or legends of Christ) did not forget to voice the threat of violence and pain in encouraging others to embrace his ideology, thus promulgating his dominance. Big hats, rising up, and making direct eye contact are all nonverbal cues that functionally communicate the same threat with the same intended outcome—submission.

As we deconstruct God's projected size and dominance behaviors, we place them within an ancient registry of human and protohuman psychology. In this way we may begin to more deeply understand the reasons why men engage in religious violence and intellectual subjugation, particularly among religions that are commanded by a dominant male god. Our primate ancestors evolved adaptations that impelled them to avoid challenging the dominance of larger, more powerful males. Human beings have inherited these tendencies and have woven primate rules for dominance and submission into the fabric of their religious cultures. Despotic men have coopted evolutionary fear structures in their alliances with dominant male gods and have used these directives to rain intimidation and suffering down on religious subordinates (or outsiders). Perhaps with evolutionary science we can better understand our complicity in unjust power arrangements and refrain from our seeming compulsion to respond to size and power with submission, particularly when such submission causes pain and injustice.

Chapter 7

MALADAPTIVE SUBMISSION TO THE GODHEAD

While offering little consolation to those suffering the mental anguish of clinical depression, scholars of evolutionary psychiatry have pointed out how low motivation, pessimism, inactivity, and other depressive symptoms very likely aided the fitness of our ancestors, chiefly by smoothing the way to submission to more powerful, more dominant, or otherwise dangerous individuals.[1] Depressive phenomena have implications for religious and political power structures, particularly for the men who claim to represent gods, whether they are god-kings, presidents, imams, popes, warlords, or millionaire evangelists, all of whom capitalize on the paralysis of depressive submission.

THE PECKING ORDER

Research in evolutionary psychopathology suggests that depression may be an adaptation. As one point of evidence, scholars in this field cite epidemiology; worldwide depression is an unusually prevalent psychiatric illness. In America, it is the most common psychiatric condition, with a lifetime prevalence of nearly 17 percent—meaning that about one out of every five Americans will suffer from a major depressive episode at least once in their lifetime.[2] We know that depression has a strong genetic component,[3] but this rate of prevalence is far greater than we would expect by chance mutation. Psychiatrist and medical anthropologist Daniel Wilson puts it this way, "Simply stated, genes common enough to have evolved by means of natural selection can have done so only by advantages conferred to lineages which have carried such genes *even if such genes now express a level of phenotypic disease.*"[4]

If it seems paradoxical to describe a recognized psychiatric disorder as adaptive, remember that nature "cares" little for how unpleasant a subjective experience may be so long as it promotes survival. The key to understanding how depression can provide a survival advantage lies in understanding the relevance to rank status. Evolutionary scientists have found that depression among social animals is linked to their respective positions in the pecking order. The term *pecking order* was first introduced in 1935 by a Norwegian zoologist, Thorleif Schjelderup-Ebbe, who used the term to describe how chickens maintain rank by the strategic use of aggression—that is, through pecking one another. He observed how a drop in rank among chickens was often followed by an apparent depressive condition:

> Deeply depressed in spirit, humble, with drooping wings and head in the dust, it is—in any rate directly upon being vanquished—overcome with paralysis, even though one cannot detect any sign of physical injury.[5]

In 1967, evolutionary psychiatrist John Price observed similar patterns in subordinated long-tailed macaques and noted their similarities with depressed psychiatric patients.[6] From these observations Price began developing a compelling evolutionary theory of depression. He proposed that depression is an ancient means of regulating rank and social competition between animals, including humans, and that it provides a temporary impetus to yield to more powerful individuals. According to Price, this *social competition hypothesis* of depression proposes that yielding has three functions:

(a) an executive function which prevents the individual from attempting to make a "comeback" by inhibiting aggressive behaviors to rivals and superiors,
(b) a communicative function which signals "no threat" to rivals, and;
(c) a facilitative function which puts the individual in a "giving up" state of mind which encourages acceptance of the outcome of competition and promotes behavior which expresses voluntary yielding.[7]

This theory is consistent with our knowledge of the evolutionary process. Getting into frays with stronger individuals is a quick way of getting removed from the gene pool. Similarly, communicating that you concede to lower rank and that you pose no challenge is an effective way of avoiding attack. The brain mediates these behaviors. In a violent world, it pays to have a brain programmed to selectively temper the motivation for behaviors with a high probability for injury or death. Humans, in contrast with macaques and chickens, may also experience depression in response to a perceived drop in status *not* precipitated by violent competition, such as losing employment, ending a marriage, or failing out of school.

There are biological processes at work here. Research shows that testosterone, for example, has implications for rank acquisition. High testosterone is associated with better mood,[8] winning competitions,[9] higher motivation to reengage after losing,[10] and higher sex drive,[11] whereas low testosterone is associated with depressed mood,[12] withdrawal from competition,[13] and low sex drive.[14] That mood, sex drive, and motivation to engage in rank competition are regulated by the same endocrine would seem to validate Price's theory. On the flip side, serotonin levels (5-HT), which play an important role in determining positive mood, are found to be higher among the higher-ranking, and the high-ranking get a boost of 5-HT upon receiving submissive displays.[15]

Price's theory helps to set the stage for understanding why religious institutions might well prosper by encouraging behaviors that mimic depressive phenomena. Evolutionary theory may also help to explain why religion is so effective at doing so—chiefly because it builds upon preexisting patterns of submission.

WORTHLESSNESS AND THE SIN OF PRIDE

A number of depressive symptoms appear to serve submission. Explaining this requires first considering the evolutionary basis for self-esteem. Evolutionary researchers have argued that self-esteem is a kind of "gauge" or "index" designed to inform adaptive goals.[16] Such a gauge is critical in social hierarchies where individuals must understand their rank status and choose social behaviors according to their rank, particularly in regard to social competition—for example, not challenging a more dangerous, higher-ranking individual or submitting to a weaker, lower-ranking one. Self-esteem then, either low or high, acts as an emotional gauge to inform which strategy to use.

An important related concept is resource holding potential (RHP), a term coined by British biologist Geoffrey Parker to describe the probability of an animal to win in an all-out confrontation.[17] Parker described how when presented with confrontation, individuals estimate their own RHP against those of rivals. This assessment has important fitness implications: wasting time and energy and risking injury by challenging more powerful individuals can be dangerous.[18] On the other hand, using estimations of RHP to inform behavioral choices can help individuals to, as behavioral ecologists John Krebs and Nicholas Davies put it, "submit to those stronger [and] challenge those weaker."[19] Such comparisons require self-representations to weigh against representations of the opponent. These representations form the basis of self-esteem, which in turn serves important motivational functions—the experience of high self-esteem motivates dominance behaviors, while low self-esteem motivates submission.

The scientific literature appears to support self-esteem as a gauge for estimating RHP and rank. For example, one study found that subjects who perceive themselves as lower-ranking were more likely to engage in submissive behaviors and to feel distress about behaving assertively.[20] Other research found that depressed individuals are more likely to make upward comparisons to those who are better performers than they are[21] and that depression is related to negative self-appraisals on dimensions of rank.[22]

Perhaps more informatively, patients with depression often view themselves as *defeated, incapable, inferior,* or *worthless*, self-perceptions compatible with submissive behavior. The current American Psychiatric Associations diagnostic manual, the DSM-5®, describes this symptom as "feelings of worthlessness or excessive or inappropriate guilt."[23] Now, like any emotion, humility—which includes feelings (or demonstrations) of lowliness, meekness, inferiority, and so forth—is experienced in gradations, with depressive worthlessness on the more extreme end. But the point here is its role in regulating social hierarchies. Human worth is almost always a relative value, established by comparison to other humans. Feelings of worthlessness in a context of social hierarchies therefore embody feeling *less worth* than *someone with more worth*, which typically translates to someone *more dominant* in some way or another. The feeling can occur after a defeat or simply as a defeat-prevention measure—after calculating RHP, which may prompt a weaker individual to choose submission rather than competition.

Nonhuman primates concede to lower RHP by shrinking down, averting eye gaze, and making other nonverbal submissive displays. As Price puts it, "The message of a submissive signal is 'You are more powerful than me' which is probably as near to 'You are wonderful' as an animal without language can get."[24] Humans make these displays as well, but they also use praiseful language to acknowledge higher rank. For example, with nobility we might use phrases like *your excellency, your highness, your majesty, your grace,* or *your eminence.* Conversely, self-abasements such as *I am pitiful, weak, lowly, meek,* or *powerless* are human means of communicating lower rank. Importantly, for humans these communiques also convey submissive states of mind—humility versus pride or low self-esteem versus high self-esteem—which tend to be good predictors of behavior (i.e., submission versus challenge).

When viewed with an evolutionary lens, the idea that humans should emphasize religious humility in relation to a superior being has great explanatory power, particularly if that being is male and fearsome. Competitions over rank are an integral part of primate social life—as we've seen, dominant males are submitted to across evolutionary history—and the familiarity of submission displays may make it emotionally easier for humans to be submissive to God. Indeed, humans exalt gods and self-abase before them to show submission just as they do with higher-ranking humans. For example, self-abasement can be shown linguistically, as we see in common refrains such as, "God is great, but we are small. God is without fault, but we are sinners." It can also be shown nonverbally, "I rose from my self-abasement, with my tunic and cloak torn, and fell on my knees with my hands spread out to the Lord my God" (Ezra 9:5) or emotionally (italics mine), "Better to be *lowly in spirit* along with the oppressed than to share plunder with the proud" (Prov. 16:19).

If we can say that humility is an experience chiefly reserved for those lower in the hierarchy, then it is no wonder that humility is awarded tremendous value across many religions and is lauded as a sign of spiritual devotion. World religions are full of references to humility and many praise its value, for example (italics mine): "And the *slaves of God* are those who walk on the earth in *humility* and calmness" (Koran 25:63). Further, as we might expect from a god based on a dominant male primate, humility in scripture is described as a means to deflect aggression (italics mine): "God *opposes* the proud, but gives grace to the humble" (James 4:6); God says, "They have humbled themselves. *I will not destroy them*" (2 Chron. 12:7).

Arguably, the antithesis of low self-worth is pride, for having pride by definition requires a sense of self-worth. Tellingly, pride is considered one of the seven deadly sins, where sin amounts to an act that *defies the rules set by a dominant male god.* Pride, like the other deadly six, has important rank and fitness implications. While extreme humility is generally appreciated by the religious aristocracy, by statesmen, and purportedly by God (not to mention males of other primate species), pride is dangerous to those in high positions and therefore not tolerated—thus subordinates among men and apes are *not* allowed to make eye contact, be loud, or stand tall. But given that God is said to be all-powerful, it seems incongruous that he should share such pedestrian intolerance for pride, although scripture suggests he does: "The LORD detests all the proud of heart. Be sure of this: They will not go unpunished" (Prov. 16:5). And not unlike powerful men, God punishes the prideful among his subordinates with aggression:

> I, the LORD, will punish the world for its evil and the wicked for their sin. I will crush the arrogance of the proud and humble the pride of the mighty. I will make people scarcer than gold—more rare than the fine gold of Ophir. (Isa. 13:11–12)

Here is an example of the severity of punishment reserved for the obstinately prideful:

> And if in spite of this you will not hearken to me, then I will chastise you again sevenfold for your sins, and I will break the pride of your power, and I will make your heavens like iron and your earth like brass; and your strength shall be spent in vain, for your land shall not yield its increase, and the trees of the land shall not yield their fruit. Then if

you walk contrary to me, and will not hearken to me, I will bring more plagues upon you, sevenfold as many as your sins. (Lev. 26:18–21).

God's displeasure with pride is a frequent theme in the Old Testament. King Herod was a particularly macabre case. Because he failed to submit to God, he was struck down by one of God's angels and consumed alive by worms (Acts 12:23).

The Koran takes a similar approach to pride: "I shall turn away from my revelations those who show pride in the world wrongfully" (Koran 7:146); "Certainly He does not love the proud ones" (Koran 16:23); "It shall be said: Enter the gates of hell to abide therein; so evil is the abode of the proud" (Koran 39:72). As a male leader, God is highly intolerant of emotional states that may result in his usurpation, and so hell is reserved for the prideful.

While showing pride invites God's wrath and showing humility may forestall it, humility can also be used as a means to solicit God's protection. We have discussed the importance of protection in landscapes swarming with dangerous predators. Often, however, protection is required not only against outside forces, but from the furies of the male god himself, which call for supplication. And so the devout may prostrate themselves before God, begging for his benevolence, as discussed in chapter 6.

Taken to extremes, religious promotion of humility can lead to self-injury, or *religious mortification*. For supplicants, mortification is another opportunity to demonstrate lower worth than their male gods (or their affiliates). For example, some Muslim Shi'ites annually flog themselves with chains and swords or cut themselves with knives to commemorate the martyrdom of Husayn ibn Ali (626–80), the grandson of Mohammed who sacrificed his life serving God in holy war. Like other submission displays, self-harm has been a means to preemptively avert divine rage. During the Middle Ages, flagellation movements spread across Europe, often as a means to curtail plagues or the prophesied apocalypse. Saint Dominic (1170–1221) whipped himself nightly with iron chains in atonement to God for the world's sinners.[25] In the modern age, the Catholic *Penitentes* of New Mexico beat bloody furrows into their backs with leather whips or submit to crucifixion—emulating Jesus' sacrifice—to appease God.

No doubt there is some exhilaration involved in such rituals, which resurrect ancient emotions from our evolutionary past. No doubt, too, these practices inundate the brain with powerful neurotransmitters, such as endorphins (chemically similar to morphine), which amplify the experience. But in a different context, such behaviors would rightfully result in forced admission to a psychiatric emergency room.

Carolyn Bynum, American historian of medieval religion, writes of extreme Catholic ascetic practices of the twelfth and thirteenth centuries, practices so severe that religious authorities eventually had to call for their abolishment. Writes Bynum, "These calls for reasonableness and interiority were clearly in part a response to such alarming austerities as wearing iron plates, mutilating one's flesh and rubbing lice into the wounds, or even jumping into ovens or hanging oneself."[26] Drinking the pus of the sores of lepers was another favored form of ascetic devotion.[27]

Losing a fight for rank can involve great pain for the loser. By self-inflicting pain, one essentially saves a dominant male god the trouble of inflicting it himself and at the same time displays submission in a very poignant way. Self-abasement is even more understandable if you truly believe that God is ready to enforce the retaliations described in scripture.

Last, in major depression, suicide and low self-esteem often coincide. Religions often canonically proscribe suicide, and research overall finds there is less suicide among the devout.[28] However, mortification of the flesh can be taken to this darkest extreme, and this ultimate act of abasement has been documented across religious history. The Circumcellions, an early fourth-century Christian sect in Romanized Africa, believed that suicide was a form of martyrdom. As such they threw themselves off cliffs or into roiling fires. They also paid other men to kill them or threatened to kill strangers if they did not agree to kill them. Often the suicides were self-punishment for sin.[29] Recently, mass suicides were branded into historical infamy with cases like that of the Peoples Temple, which in 1978 saw 918 Americans drink cyanide in Guyana at the order of their dominant male leader Jim Jones. Another example is the Solar Temple, where 74 cult members committed suicide; some had donated over $1 million to their male leader, Joseph Di Mambro, prior to taking their lives. There was also Heaven's Gate, in which some male members voluntarily castrated themselves before committing suicide in order to board the UFO that was supposed to take them to a different dimension.[30]

I hold as self-evident that healthy people don't castrate themselves, drink the pus of lepers, rub lice in their wounds, jump into ovens, self-mutilate, or commit suicide. Even in a religious context, these behaviors seem clinically pathological and many are maladaptive in evolutionary terms—killing or castrating yourself removes you from the gene pool and the other self-harming behaviors risk your presence there. And yet for those who have been encouraged to believe they are worthless before God, or who live in fear of his judgment, such behaviors may be a way of showing pitiable submission before a fearsome and all-powerful ruler.

ANHEDONIA

The next depressive symptom of particular relevance for the religious is anhedonia. Defined as the loss of ability to experience pleasure, anhedonia commonly affects the appetitive drives for sex and eating. Symptoms are evinced across species, often following social defeats,[31] and serve as a motivation to surrender sex and food to more powerful individuals. Accordingly, the most basic of evolutionary drives toward sex and food are also featured among the deadly sins—as *lust* and *gluttony*. Establishing these drives as sin embeds in religious canon the primordial pattern of relinquishing resources to a powerful male.

Sex and the Sin of Lust

Clinical and evolutionary science researchers Paul Gilbert and Michael McGuire point out that animals influence each other's emotions and behaviors by sending and receiving signals.[32] Regarding mate competition, these signals have important survival implications. Brimming with confidence in the pursuit of the same sexual resources as more dominant individuals can prove dangerous[33]—as we have seen, male macaques, chimpanzees, kings, warlords, and clergy are all prone to wreaking violence on their sexual competitors. On the other hand, showing sexual restraint (for instance, steering clear of the dominant male's females) may be a way to avert unnecessary or even mortal attacks.

One way to communicate sexual restraint to the higher-ranking is to demonstrate sexual shame (or guilt, considered a less public experience of shame). As it turns out, research finds that shame is related to depression, feelings of inferiority, and submissive behaviors.[34] Hence displaying shame—averting the eyes, lowering the head, and hiding—is an appeasement display designed to communicate submission.[35] Dominant members usually have more power to create feelings of shame in the lower-ranking than vice versa and use shaming to promote their own self-interests.[36] These dynamics of shame, rank, power, and sexual repression are highly visible in depictions of God's interactions with his subordinates.

For example, in the book of Jeremiah, God (who demands exclusivity), berates Israel for her sexual indiscretions:

> "If a man divorces his wife and she leaves him and marries another man, should he return to her again? Would not the land be completely defiled? But you have lived as a prostitute with many lovers—would you now return to me?" declares the LORD. "Look up to the barren heights and see. Is there any place where you have not been ravished? By the roadside you sat waiting for lovers, sat like a nomad in the desert. You have defiled the land with your prostitution and wickedness." (Jer. 3:1–2)

For this act of whoring with foreign gods, God sends drought to Israel, "Therefore the showers have been withheld, and no spring rains have fallen. Yet you have the brazen look of a prostitute; you refuse to blush with shame" (Jer. 3:3). This passage seems to imply that shame is the expected response to cheating on God. Further, it shows that God enacts dangerous punishments for infidelity. Eventually God forgives Israel, but demands Israel's expressed guilt:

> "I will not be angry forever. Only acknowledge your guilt—you have rebelled against the LORD your God, you have scattered your favors to foreign gods under every spreading tree, and have not obeyed me," declares the LORD. (Jer. 3:12–13).

Similarly, in the book of Romans Paul writes how those who worshipped idols fell out of God's favor and in doing so were left to their shameful sexual acts: "So God abandoned them to do whatever shameful things their hearts desired. As a result, they did vile and degrading things with each other's bodies" (1:24). Failing to show appropriate shame, the sinners incurred God's wrath: "But because of your stubbornness and your unrepentant heart, you are storing up wrath against yourself for the day of God's wrath, when his righteous judgment will be revealed. God will repay each person according to what they have done" (2:5–6).

With sex framed as a shameful infraction against God, it is no wonder that sexual shame and guilt seem to be amplified in religions. One large survey of more than 9,500 participants conducted among Americans found that the devoutly religious experienced more sexual guilt than atheists.[37] In this study, Mormons rated highest on sexual guilt, followed closely by Jehovah's Witnesses, Pentecostals, Seventh-day Adventists, and Baptists, while agnostics and atheists were among the lowest. Further, of people raised in highly religious homes, 22.5 percent reported being shamed or ridiculed for masturbating, compared with only 5.5 percent of those brought up in the least religious homes. Some of the former reported being told as children that they would burn in hell for masturbating, while others were beaten for engaging in self-pleasure. A whopping 79.9 percent of respondents raised in highly religious homes reported feeling guilty about a specific sexual act or desire, compared to 26.3 percent of respondents raised in secular homes. Further, the study found that religious observers who later became atheists noted a drastic improvement in sexual satisfaction.[38]

Much of religion's sexual repression emerges from the notion that certain sexual behaviors displease God. He has been described in the Abrahamic faiths as having personal distaste for extramarital sex, homosexuality, prostitution, oral sex, anal sex, masturbation, revealing clothing, and even sexual thoughts. One tactic prescribed in Christian theology is to force forbidden thoughts from awareness. This tactic is based on the notion that merely thinking of a sexual act is the same as committing it, for example, "You have heard that it was said, 'You shall not commit adultery.' But I tell you that anyone who looks at a woman lustfully has already committed adultery with her in his heart" (Matt. 5:27–28).

In some sects, God proscribes all but the essential sexual behavior that propagates his devotees across the generations (which he can't afford to dispense with). The devout are expected to abandon sex as a sign of submission to God. Abandoning sex to appease dominant males is an arrangement familiar to primates.

In an evolutionary context, temporarily abandoning sex can be adaptive as an act of concession to a more powerful group member (i.e., live to have sex another day), while in a clinical sense it can be pathological. Like physical pain, we are programmed to steer clear of painful emotions such as shame and guilt, which we accomplish by avoiding behaviors that generate them. In this light I argue that these emotional states can be pathological simply because they impede the experience of pleasure. In a world with such a vast menu of possible anguishes to suffer (e.g., hunger, thirst, extreme temperatures, disease, physical injury, losing a loved one, and so on), giving up sexual pleasure, or adulterating it with guilt, is an act of asceticism destined to create a depressive reaction. Indeed *not engaging* in pleasurable activities can maintain or aggravate depression, and treatment for depression often involves the systematic performance of pleasurable activities (including sex), which works exceptionally well at treating the disorder.

Like other forms of religious abasement, displays of sexual renunciation can be taken to extremes. Genital mutilation has been used for millennia to show loyalty to God and lends itself easily to an evolutionary lens. In Genesis, chapter 17, we have Abraham's renowned covenant with God brokered by a powerful act of submission—Abraham circumcises himself. While Abraham's circumcision ends up being a hallowed, central credo of Judaism, it is equally an act of alliance-making that mirrors the alliances primates have always had with dominant males. Here we have a lesser male (Abraham) showing submission to a dominant male (God) through a display of sexual subservience (by harming his own genitalia). In exchange Abraham gains assistance in acquiring territory (the land of Canaan), high rank ("I will make kings come from you"), and measured reproductive success (Abraham's son Isaac). In another instance (Exod. 4:24–26) God intends to kill Moses, but his wife, Zipporah, saves him by circumcising their son with a flint knife and casting the bloody foreskin at Moses's feet. Satisfied with this act of acquiescence, God spares Moses's life. We have already heard how enraged primates dominate by attacking genitalia, which ends up being an effective way to eliminate sexual rivals. Here God is satisfied with a symbolic act of genital mutilation as a means to show Moses's concession to lower (sexual) rank. Conceivably God could have demanded Moses's earlobe or lip or hair, but predictably, he preferred the foreskin.

It is probably not incidental that circumcision results in less sexual sensitivity due to callousing of the nerve-ending-rich penile glans; circumcision, in effect, produces less sexual pleasure, something that one can easily imagine dominant males wishing on would-be sexual competitors. The foreskin itself probably served some evolutionary function rooted in mate competition. One researcher points out that among monogamous primates, males have no foreskin, but among promiscuous primates, males have foreskins and other "elaborations including spines, plungers, labile scoops, flexible ridges and other distal structures" designed to "increase the probability that a male's sperm will achieve fertilization and decrease this probability for rival sperm."[39] In short, if the foreskin is an evolutionary adaptation for increasing the ability to compete sexually with other males, cutting it off not only dulls sexual sensitivity but may also make one less sexually competitive.

Another example of religious genital mutilation is described among the ancient Aztecs, who pierced their penises with cactus thorns or stingray barbs to honor their male gods. If a young man fainted during the ritual, it was seen as proof that he failed to keep his virginity.[40] The Mayans had a similar ritual in which men let blood from their penises to honor their male god Tohil.[41]

Castration remains perhaps the most extreme example of this kind. Castration for God, needless to say, is the permanent ceding of reproductive capacity. The ancient Zapotecs (rivals of the Aztecs) castrated their elite children to prepare them to serve their gods as neophyte priests.[42] Some early Christian sects advocated castration as a means to sacrifice one's will to God. Initially, the physical act of castration was a nagging problem for the Church, as the extreme nature of the sacrifice made Christianity less appealing to potential male converts. The Church formally outlawed this behavior at the First Council of Nicaea in 325 CE.[43]

A later example comes from the Skoptsy, a Christian sect in Imperial Russia whose members believed that Adam and Eve grafted pieces of fruit from the tree of knowledge to their bodies to form breasts and testicles. To counter this "original sin," they performed mastectomies and castrations. There were two kinds of castrations, known as *minor* and *royal "seals."* While the minor seals only removed testicles, the royal seals also removed the penis. Initially the Skoptsy pressed red hot irons into the scrotum, thus destroying the testicles in a process known as "fiery baptism." Other techniques involved knives and razors (while chanting "Christ is risen"), or twisting the testicle until the seminal vesicles were destroyed, causing the testes to fall inside the scrotum and eventually atrophy.[44]

Female genital mutilation is another religious practice in which we can see the designs of male sexual control. Skoptsy women had their nipples or entire breasts removed and sometimes the labia majora, the labia minora, and the clitoris. The Dogon tribe of Mali offers another example. The Dogon believed that their male god Amma created the earth with a lump of clay. With it he made a female body with a sex organ in the shape of an ant mound and a protrusion in the form of a termite hill (the clitoris). When he tried to lie down and copulate with the earth, the symbolic clitoris got in the way so he excised it.[45] Today, Dogon girls have their clitorises and labia ritually excised by the resident blacksmith. Regardless of the surrounding mythology, removing the clitoris deadens female sexual pleasure. This severe act of female sexual control has reached epidemic proportions. The World Health Organization estimates that some 140,000,000 women and girls worldwide are suffering the effects of female genital mutilation.[46]

Female genital mutilation is practiced in several Islamic cultures. While some Islamic sects have issued fatwas against the practice, others have supported it as a means to ensure chastity and ultimately sexual loyalty to Muslim husbands. However, as is typical, the wishes of men and their male gods are often blurred and so the capricious female lust so threatening to men's reproductive fitness achieves a certain parallel with the male lust considered so treacherous to a dominant god. Recently Iranian Ayatollah Kazem Sedighi warned his followers that when women dress inappropriately (eschewing traditional Islamic drapery) this causes extramarital affairs, which in turn cause earthquakes.[47] One can only assume that the earthquakes are considered an expression of God's displeasure.

Food and the Sin of Gluttony

> "And put a knife to thy throat, if thou be a man given to appetite." —Proverbs 23:2

In addition to a loss of appetite for sex, depression can also involve a loss of appetite for food. As a means of encouraging the surrender of food to more powerful individuals, this symptom is consistent with the social-competition hypothesis of depression. Surrendering food has its roots in biological evolution, but continues to be played out in religious ritual.

Food is the primary focus of resource competition in the biologic world. Utilization of every other resource requires food. Skulking and hunch-shouldered, subordinates across the animal kingdom will surrender food claimed by a dominant, satisfying themselves with whatever scraps remain. High-ranking mammals, such as lions, wolves, and hyenas, are often seen displacing lower-ranking members from a kill. Similarly, higher-ranking primates often displace lower-ranking individuals from key feeding sites. For instance, higher-ranking male chimpanzees often feed higher in the canopy of trees where fruit is more abundant and sugar content is higher. When confrontations arise, lower-ranking members withdraw, either by descending to less desired parts of the tree or by leaving the tree altogether.[48] High-ranking chimpanzees also steal meat more often and are given meat more often by lower-ranking males.[49] When presented with an unwinnable confrontation over food, a sensible response is to relinquish that resource rather than risking injury or death.

To put this in religious context, let us recall again the Genesis story, a story central to Judeo-Christian mythology in which God banishes Adam from the Garden of Eden for eating fruit from the tree of life. The conventional symbolism for the forbidden fruit is knowledge, a topic to which we will shortly return. However, the significance of this story in evolutionary terms is so obvious that it becomes almost painfully simplistic: the tale describes a dominant male (God) expelling a subordinate (Adam) from his territory (the Garden of Eden) for daring to intrude on his forbidden food resource (fruit of the tree of knowledge). Tertullian (160–225 CE), an important early Christian writer who has been called the "the founder of Western theology," labeled this first sin as *gluttony* (a deadly sin).[50] Carolyn Bynum cites another influential early fifth-century Christian writer, Saint Nilus the Elder, who reasoned: "It was the desire for food that spawned disobedience; it was pleasure of taste that drove us from Paradise. Luxury in food delights the gullet, but it breeds the worm of license that sleepeth not."[51] According to many Christian denominations, this original sin set in motion an everlasting cascade of pain for every succeeding generation of human and non-human animal. The sin even made eternal life impossible without the intervention of God. As described in the Bible, God comes down hard on Adam and Eve, as we would expect from a dominant male:

> To the woman he said, "I will make your pains in childbearing very severe; with painful labor you will give birth to children. Your desire will be for your husband, and he will rule over you."
>
> To Adam he said, "Because you listened to your wife and ate fruit from the tree about which I commanded you, 'You must not eat from it,' Cursed is the ground because of you; through painful toil you will eat food from it all the days of your life. It will produce thorns and thistles for you, and you will eat the plants of the field. By the sweat of your brow you will eat your food until you return to the ground, since from it you were taken; for dust you are and to dust you will return." (Gen. 3:16–19).

In this most primeval biblical landscape, God enacts the most primal conflict—violent competition over food. When his control of the food source is challenged, he displaces his challengers. He then forces women into pain and subservience and men into a lifetime of toil. In returning them to dust he gives to exquisitely self-aware humankind the annihilation of self it so desperately fears and forces humans to rely on him for protection against death. The story as told would suggest that food is immensely important to God, as is surrendering it to him. This is despite the fact that God is described as an everlasting, omnipotent being—one that should be indifferent to food, not needing it for survival.

Yet religious practice often emphasizes surrendering food resources; the ritual sacrifice of food to gods is common across world religions. Often such sacrifices are made to secure the favor of gods or to receive their blessings. As demanded in the Old Testament:

> Make an altar of earth for me and sacrifice on it your burnt offerings and fellowship offerings, your sheep and goats and your cattle. Wherever I cause my name to be honored, I will come to you and bless you. (Exod. 20:24)

Knowing that this dominant male god doesn't physically require the surrendered food of humans, one might question the purpose of such sacrifices. And yet his hunger is large, requiring an extravagant menu. The biblical god is banqueted a vast array of foods: bulls, rams, lambs, grain offerings, and drink (Ezra 7:17); doves and pigeons (Luke 2:24); sheep, goats, and cattle (2 Chron. 7:5); prized internal organs and fat (Lev. 3:3); oxen (Num. 7:17); and olive oil (Lev. 9:4). It is ironic how much Christians have criticized the indulgences of the Roman aristocracy in the era of Christ, given the luxuriant appetites of their God.

Fasting is another means of surrendering food resources, and the Bible abounds with examples of this gesture of appeasement. Now, numerous other reasons for fasting exist. Today people fast to diet, to self-cleanse, to prepare for medical procedures, or to make political statements. However, examples of fasting in the Bible are almost always meant to show submission, the dutiful fast to avert God's wrath or to plead for his favor or protection.

For instance, the prophet Joel calls for a fast to appease God's wrath, which on this occasion was a plague of locusts sent to torment Joel's people, "That is why the LORD says, 'Turn to me now, while there is time. Give me your hearts. Come with fasting, weeping, and mourning.' ... Who knows? Perhaps He will give you a reprieve, sending you a blessing instead of this curse" (Joel 2:12, 14). This passage explicitly makes the connection between depressive states (e.g., weeping and mourning), submission, and surrendering food. When God threatened to destroy the city of Nineveh for worshipping idols, the people fasted and wore sackcloth as self-punishment. To appease God the people went so far as to force their animals to fast and wear sackcloth as well. God responded by sparing the city (Jon. 3:6–10). Fasting is also undertaken as a means to appease God's wrath for disloyalty, "Then Ezra withdrew from before the house of God and went to the room of Jehohanan son of Eliashib. While he was there, he ate no food and drank no water, because he continued to mourn over the unfaithfulness of the exiles" (Ezra 10:6). This occurs again in Deuteronomy, and the gesture appears effective, "Then once again I fell prostrate before the LORD for forty days and forty nights; I ate no bread and drank no water, because of all the sin you had committed, doing what was evil in the LORD's sight and so arousing his anger. I feared the anger and wrath of the LORD, for he was angry enough with you to destroy you. But again the LORD listened to me" (Deut. 9:18–19).

The pious also give up food to ensure God's protection:

- In exchange for safety from bandits: "There, by the Ahava Canal, I proclaimed a fast, so that we might humble ourselves before our God and ask him for a safe journey for us and our children, with all our possessions." (Ezra 8:21).
- To curtail the destruction of Jerusalem: "I, Daniel, understood from the Scriptures, according to the word of the LORD given to Jeremiah the prophet, that the desolation of Jerusalem would last seventy years. So I turned to the Lord God and pleaded with him in prayer and petition, in fasting, and in sackcloth and ashes. (Dan. 9:2–3)
- For protection against illness: "Yet when they were ill, I put on sackcloth and humbled myself with fasting." (Ps. 35:13)
- And for victory in the dangerous business of warfare: Then all the Israelites, the whole army, went up to Bethel, and there they sat weeping before the LORD. They fasted that day until evening and presented burnt offerings and fellowship offerings to the LORD. And the Israelites inquired of the LORD. . . . They asked, "Shall we go up again to fight against the Benjamites, our fellow Israelites, or not?" The LORD responded, "Go, for tomorrow I will give them into your hands." (Judg. 20:26–28)

Virtually every Abrahamic sect observes fasting rituals, many involving penance for sins. Fasting during the widely practiced traditions of Jewish Yom Kippur, Christian Lent, and Islamic Ramadan is based on penitence and submission across the three great Abrahamic faiths.

Given its relationship to depression, it is understandable that submissive fasting can manifest in self-hatred and self-harm. During the European Middle Ages, a period stirring with asceticism, forms of fasting were widely practiced among Christians, often in extreme forms. The devout fasted as penitence for Adam's original sin. Saint Francis of Assisi wrote that, "I have no greater enemy than my body," and taught, "We should feel hatred towards our body for its vices and sinning!"[52] For Saint Francis this belief involved starving the body as a means of atonement—along with self-flagellation and referring to his body as "brother donkey."[53] Expanding on the beast of burden metaphor, Saint Francis of Bonaventure wrote that the body (italics mine) "should be weighed down by hard work, often scourged with the whip, and *nourished with poor fodder.*"[54] Saint Catherine of Siena (1347–1380) and Saint Teresa of Avila (1515–1582), the two great female doctors of the Church forced themselves to vomit, reportedly by sliding plant stems and branches down their throats. According to legend many ascetics were able to live entirely without food in their efforts to atone to God, including Benevenuta of Bojanis, Elsbeth Achlerin, Saint Lidwina of Schiedam, and the Swiss national saint, Nicholas of Flüe.[55] Needless to say, scores of ascetics died desperately trying to slide closer to God's side by surrendering food.

While death by self-starvation hit a frenzied peak in the Middle Ages, it is known to occur in the modern age as well. Just as self-mutilating for God can be considered pathological, so can starving oneself to death. The behavior speaks to complete diminution of self-worth in order to appease a more powerful being. The Christian Church has elaborated upon the practices of fasting and celibacy and raised them to the heights of religious ecstasy. But surrendering food or sex to soothe the rages of a dominant being are behaviors that reach back to the primordial savannah.

DIMINISHED ABILITY TO THINK

Another depressive symptom described in the DSM-5® is the "diminished ability to think or concentrate, or indecisiveness."[56] While deficits of this kind are very often maladaptive, they too may serve an inhibitory function that in specific (temporary) circumstances may have survival advantages. While indecisive, for example, one may be less likely to suddenly act on potentially harmful impulses such as challenging more powerful or higher-ranking group individuals. Supporting the link between indecision, inhibition, and rank, research on depression in humans demonstrates that cognitive deficits are strongly predicted by psychomotor retardation (diminished ability to move or act),[57] and animal research finds that social defeats cause both cognitive impairment and depression.[58]

Evolutionary biologists Paul Watson and Paul Andrews point out that the varied cognitive deficits seen in depression are generally abstract and nonsocial, but that depressed persons tend to be hyper-focused on social information and actually better at some socially oriented cognitive tasks.[59] They argue that this differential performance may have provided a fitness advantage—chiefly by allowing the depressed individual to focus on social problems, while reducing distraction by other kinds of mental tasks. Also, in calculating RHP, accuracy is important; a gross overestimation of one's own fighting ability compared to another's could be prove deadly. The authors draw attention to the fact that those who are depressed tend to be more realistic about their capabilities than those who are not depressed, which can be a valuable asset in social decision-making around conflict. Nonetheless, it remains largely unclear to what extent the cognitive limitations seen in depression may have served an adaptive purpose—related to inhibition, social rank, or diverting attention to socially oriented mental tasks—and to what extent they represent more of the "system failure" of a true pathology. More research is needed on this subject.

However, here I suggest that cognitive impairment works well as a metaphor for a more mundane lack of knowledge. Insofar as both a sharp mind and knowledge can be used to question the legitimacy of power or to plan actions against the powerful, the metaphor seems fitting. Ignorance, in other words, may function as effectively as impairment in keeping one out of unwinnable conflicts.

It is interesting to remember that, according to the Judeo-Christian story, it was eating the apple of the tree of knowledge that doomed humankind to penury and pain, forced us to struggle for food to survive, and made us mortal. The apple was literally a piece of food, and common interpretations suggest it may also represent sexuality, but the tree of knowledge from which it was plucked gave to humankind a level of understanding reserved only for god(s):

> Then the LORD God said, "Look, the human beings have become like us, knowing both good and evil. What if they reach out, take fruit from the tree of life, and eat it? Then they will live forever!" So the LORD God banished them from the Garden of Eden. (Gen. 3:22–23)

We are left with a dominant male banishing from his territory those subordinates who have acquired knowledge with the potential to threaten his status (here as an immortal). Knowledge as threat is recapitulated across the ages among Christian thinkers. In John Milton's epic poem *Paradise Lost*—considered one of the greatest works of the English literature—he tells of the fall of humanity from grace. He writes of the threat of knowledge:

> But Knowledge is as food, and needs no less Her Temperance over Appetite, to know In measure what the mind may well contain, Oppresses else with Surfeit, and soon turns Wisdom to Folly, as Nourishment to Wind.[60]

Christian thinkers have long been wary of knowledge. Thomas Aquinas is reported to have said, "I am a man of one book [the Bible]," seeming to imply that all others might be suspect. The great reformer Martin Luther assailed reason in a fervor:

> Reason is the Devil's greatest whore; by nature and manner of being she is a noxious whore; she is a prostitute, the Devil's appointed whore; whore eaten by scab and leprosy who ought to be trodden under foot and destroyed, she and her wisdom. . . . Throw dung in her face to make her ugly. She is and she ought to be drowned in baptism. . . . She would deserve, the wretch, to be banished to the filthiest place in the house, to the closets.[61]

Non-Christian knowledge was long considered sinful for Christians. This notion of sinful knowledge pervades even contemporary Christian thought, where the term *worldliness* is seen as akin to *sinful knowledge* rather than being a quality to admire. The fear is, as ever, that knowledge will lead to unbelief, and so Holy Scripture must take priority over other sources of information.

Just as cognitive impairment may prevent a depressed individual from acting against more powerful competitors, religion may impair the masses, effectively preventing them from acting against the prevailing power structure. One of the most effective ways to create cognitive impairment is to simply strangle the flow of information. Unwanted books may be put to flame, censored, or banned outright. In 1559, the Catholic Church issued the *Index Librorum Prohibitorum,* an index of prohibited books that blocked from the mind of the populace certain scientific texts, as well as books considered immoral or translations of the Bible into the "common tongue" (as noted in chapter 6, outside religious texts and laypersons reading the Bible are both considered dangerous). There have been numerous iterations of the index over the centuries, although it was finally abolished in 1966 by Pope Paul VI. Nevertheless, Canon Law still reserves the right to censor certain books or to grant *imprimatur,* or "let it be printed."

The absence of religious knowledge endures among Christians, which is not surprising considering the tradition of prohibition and censorship in Christianity, including of the Bible. The Pew Research Center's Forum on Religion and Public Life conducted a large survey of 3,412 Americans and found that atheists and agnostics scored highest on a measure of religious knowledge, surpassing Jews, Mormons, Catholics, and various Protestant sects (including Evangelicals).[62] These results raise many important questions. One is: If the dutifully religious had more knowledge of religion, would they so readily engage in religious submission?

While Christians have achieved legendary acts of censorship, some contemporary Islamist groups have taken up the charge in medieval form. One example is the case of Theodoor van Gogh, a Dutch filmmaker and writer who was killed in 2004 by an extremist Dutch-Moroccan Muslim for producing a controversial film about the treatment of women in Islam.[63] The assassination suggests that ideas about equal rights for women produce strong, evolved threat signals in cultures focused on female sexual control.

Another example of Islamic censorship was the controversy prompted by the Danish newspaper *Jyllands-Posten* when the paper published editorials depicting the prophet Mohammed in cartoon form. The newspaper stated that the publication was an attempt to stimulate discussion about self-censorship in Islam, where depictions of Mohammed are considered heretical. Violent Islamist protests ignited across the Muslim world. The Danish embassy in Pakistan was bombed and Danish embassies in Syria, Iran, and Lebanon were set on fire. Hundreds of deaths around the globe resulted from reactions to the cartoon.[64]

According to watchdog organizations concerned with journalistic freedom and free speech, Islamic nations today predominate among the top ten most censored countries in the world.[65] Many Islamic countries show rigid religious intolerance and censor any line of thinking antithetical to the tenets of the Koran or Islamic culture. Some have kept their populace absolutely ignorant of the outside world. But Islam wasn't always like this. It once was the world's most potent advocate for intellectual freedom, a wellspring of poetry, art, music, medicine, philosophy, mathematics, and technology, at a time when Christian Europe was fumbling through the Dark Ages. Many historians argue that Islamic peoples, having saved and then reintroduced the classics of Greek antiquity, were instrumental to the European Renaissance. There were huge centers of learning across the Islamic world that welcomed all knowledge, such as the House of Knowledge in Baghdad, which translated manuscripts from the far corners of the globe and was considered the richest center for intellectual growth in the world. In this age, religious tolerance was the mark of Islam. Islamist leaders, instead of spending their time censoring knowledge, amassed vast personal libraries. Al-Ma'mun, a Caliph of Baghdad, arranged for foreign rulers to pay tribute in the form of books from their libraries[66] and is said to have paid its weight in gold for the translation of any book yet to grace his vast collection. But more recently, driven by religious fervor, many Islamic societies have followed the lead of their Christian predecessors, punishing intellectual curiosity with human slaughter.

How has this happened? Some scholars have pointed to economic privation, while others have pointed to lack of education—both hypotheses are not without holes. For instance, many suicide bombers, it turns out, emerge from the wealthy and educated classes.[67] We turn then to the Koran and the multitude of verses that threaten unbelievers. For example: "They [unbelievers] shall be held up to shame in this world and sternly punished in the hereafter" (Koran 2:114); "Those who deny Our revelation We will burn in fire" (Koran 4:56).

But violence may ultimately be a less effective strategy than simply choking off the flow of information. An interesting comparison has been made between Arabia and Spain (a territory once dominated by Islamic rulers): the number of books that the whole Arabic world has translated into Arabic since the ninth century equals the number of books Spain translates into Spanish in one year.[68] This deprivation has occurred while the uneducated Islamic masses are being funneled into the "fundamentalist machinery of the madrassas," to borrow from neuroscientist and secular advocate Sam Harris.[69] Suffice it to say, the flow of information to these schools is asthmatically narrow. Notably missing are the writings of the great free-thinkers of world history. In such an environment, where is one to turn for alternate perspectives?

Whether by interdiction or violence, it requires effort to limit the flow of knowledge and control the ideological landscape. But the potential benefits are immense, particularly in terms of controlling material resources. Following the patterns of dominant apes, men in positions of religious power have used the threat of a dominant male god (represented as the punishments of purgatory or hell) to trigger submissive responses and collect surrendered resources from a subordinate populace. For instance, lasting well into the sixteenth century, the Roman Catholic Church sold *indulgences*, payments made in exchange for remittance of sin, which were often paid to avoid hell or to lessen time in purgatory. This trade was by nature a corrupt enterprise, often serving personal gain or financing special projects such as the Crusades or the construction of lavish cathedrals. Perhaps the most extravagant of these was the Vatican, a gilded, frescoed castle that rivals the palaces of kings and sultans across the ages—built while pious Catholics around the globe lived in poverty.

Heavenly inducements have long been an established strategy of the Catholic Church for amassing power. Lands and riches have been historically ceded to the Church by those eager to secure a place in Heaven. Will Durant writes of how during the Middle Ages, secular landowners were required to pay *tithes*, a tenth of their income (or produce) to the local Church. These tithes were collected by priests whose job it was to "curse for his tithes," or, to excommunicate anyone fabricating or evading his returns.[70] Landowners were forced to will the Church money upon their death, and the drawing of all wills was decreed to be overseen by priests. Of the draw for such behaviors Durant writes, "Gifts or legacies to the Church were held to be the most dependable means of telescoping the pains of purgatory."[71] As a result of bartering with the souls of the populace, the clergy became fat, gorging on rich foods and drinking good wines. They also amassed property, and serfs—a kind of modified slave bound to serve the lord of the manor in his fields, forests, and mines. A single Cathedral, monastery, or nunnery might own thousands of manors. Durant describes the scope of wealth acquired by these means:

> The bishop of Langres owned the whole county; the abbey of St. Martin of Tours ruled over 20,000 serfs; the Bishop of Bologna held 2000 manors; so did the abbey of Lorsch; the abbey of Las Huelgas, in Spain, held sixty-four townships.[72]

Thus the Catholic Church succeeded in manipulating fears of damnation to amass power and wealth around the globe at the expense of large masses of subordinated Church followers. I see the susceptibility to scams such as these to be evidence of cognitive impairment among the laity, with the cause being a canon that uses one evolved mechanism—submission—to circumvent another—cheater detection.

One would think that chicanery of this kind could only be successful among a populace of uninformed medieval minds. But today, American television teems with millionaire televangelists using high-pressure sales tactics to acquire profane fortunes, which they use to influence politicians and build multimillion-dollar churches on expansive compounds— churches that in America are not required to pay taxes. Using the direct exchange strategies of their medieval predecessors—money for good fortune, for health, for salvation—these men (and occasional women) coerce viewers into emptying their bank accounts with unabashed refrains like, "Jesus wants you to send in your money," in the hope of bigger and better blessings to come. They stage healings where convulsing, babbling actors are cured instantly with the thrust of a hand. Many of these televangelists have amassed stunning fortunes—mansions with private air strips, expensive cars, wives bedecked with huge diamonds—off the grocery money of mesmerized grandmothers hoping for an easier life in the afterworld.

So it is that when religious submission is disassembled, its similarities to depressive symptomology are incontrovertible, with clear implications for impaired functioning, personal suffering, and danger to self—the marks of clinical pathology. But religions mask over the symptoms of depressive submission, in some cases purporting to exchange the resulting suffering for ecstatic union with God. This experience, touching as it does on our evolved desire to affiliate with dominant males, has a strong pull, and even the most extreme cases of abasement and self-injury can spread to pandemic proportions, such as with the ascetic practices of the Middle Ages. However, with knowledge now at our disposal to explain the evolutionary patterns in which our religious cultures are embedded, we can generate enough reflective distance to make healthier decisions about behaviors that generate self-harm or that serve despotic power structures, both recognizable consequences of religious submission.

Chapter 8

THE FEARSOME REPUTATIONS OF APES, MEN, AND GODS

"The blow of a whip raises a welt, but the blow of the tongue crushes the bones." —Ecclesiasticus 28:17

THE ORIGINS OF REPUTATION

Chimpanzees can drag, kick, hit, or jump on their opponents, breaking bones or causing internal bleeding. They can also bite with teeth that can puncture meat, rip off faces, clip off digits, or tear off genitalia. Often, being of lower rank means being the recipient of the largest share of violence of these kinds. As we have seen, the privilege of male dominance imparts great reward, notably more females and more food. But it also means *not suffering* the pains of being dominated.

Thus, while achieving dominance comes with many advantages, losing dominance is a dreaded affair. When not killed outright, a deposed alpha male is frequently beaten down by all those he once subordinated and is sometimes forced to the lowest rank position in the entire hierarchy. Occasionally these former alphas are forced to live on the group periphery where they are more vulnerable to attacks by outsiders. These patterns, so conspicuous among nonhuman primates, are just as evident in reviewing the fate of dominant men who fall in rank. Deposed male leaders have been historically exiled, tortured, or publicly executed (sometimes along with their entire family). The following are a few familiar examples: Julius Caesar (Rome), Nicolae Ceaușescu (Romania), Saddam Hussein (Iraq), and Moammar Gadhafi (Libya). Given the high costs of being deposed and the significant rewards of the alpha position, alpha males often have the greatest incentive of all to maintain dominance.

In this endeavor, posturing is an important tool. Like other primates, alpha male chimpanzees posture in a manner designed to inspire awe—bristling hair, baring teeth, screaming, swinging through branches, hurling rocks, smashing sticks, drumming on roots, and even bashing clangorously on empty kerosene cans stolen from primatologists' camps.[1] These displays are by nature public, signifying dominance rank to the group at large and reminding subordinates of expected behaviors such as submission, appeasement, loyalty, and surrendering resources. If the alpha's displays are sufficiently intimidating, they may amplify the perceived cost of challenging him. In short, displays of dominance are strategies that serve to keep the power of the alpha fresh in the minds of his group members.

Here, the phenomenon in chimpanzees is described by two eminent primatologists, Richard Wrangham and Dale Peterson:

> (Nervous alpha males get up early, and often wake others in their overeager charging displays.) And all these behaviors come not from a drive to be violent for its own sake, but from a set of emotions that, when people show them, are labeled "pride" or, more negatively, "arrogance."[2]

The key insight here is that dominance displays include information related to an immense struggle for survival in a complex social environment. Being known for one's ferocity is a vital tool for maintaining dominance. Posturing, bristling, and thrashing kerosene cans are all behaviors employed to build a male's reputation and are therefore practiced with great intensity.

A reputation for dominance involves more than just posturing. It may also be maintained by random acts of unprovoked aggression. At other times, aggression is not random but reactionary. If an alpha perceives a challenge to his authority—or even a hint of challenge—he is likely to attack the offender and quickly put him in his place. Other group members, witnessing the violence, are reminded of the costs of insubordination. In cases where an alliance rises against the alpha, the allies of would-be usurpers will also be attacked as punishment.[3] Such attacks not only set the precedent that insubordination will not be tolerated but also that allying against the alpha will be met with violent retribution. Once this perception is established, it sits in the memory of other group members, and the alpha wins a reputation as one that crushes not only insubordination but also any alliances formed against him. This is of particular importance because it is often coalitions that accomplish the violent dethroning of alphas.

The pressure to establish dominance is not always a function of being male. In some instances female competition is also defined by violence. Female macaque monkeys, for example, are known for their aggression. They are also unique among primates in that they engage in line warfare with females from other groups. These wars are fought with exceeding ferocity and for a good reason. Dario Maestripieri discusses potential privileges and costs of warfare:

> When individuals or groups meet and fight for the first time, they are fighting not for a single meal, but for power, and power is not about this or that meal, but potentially every other meal for the rest of their lives.[4]

The point I wish to emphasize here is that the outcome of status competition has a lasting impact on social animals that have a keen memory of where other members lie on the social hierarchy and of the implicit rules of their respective positions. In evolutionary terms, reputation can be operationalized as the *memory of rank status*. Maestripieri calls attention to the fact that the implications of competition are particularly enduring in societies in which dominance hierarchies are stable.[5]

Some human hierarchies are stable and last for many generations, while some are constantly shifting. Even so, the play for dominance is ubiquitous; to win or maintain dominance, humans posture and attack on playgrounds, in prisons, on sports fields, in urban neighborhoods, in the corporate world, in academia, in religion, and in nearly every other social venue imaginable. As in nonhuman primates, posturing is witnessed by other humans who respond either by showing deference or by challenging the aggressor for position.

But while human and nonhuman primates use physical displays of power to establish reputation, humans also use their evolved capacity[6] for language. It follows then that humans devote a great deal of language to convey information about things like competition, rank status, and power. Language operates in a unique fashion because it allows the communication of social information to span time and geographic space, thus dispersing across larger networks in a more enduring way. The written word lengthens the reach of reputation even further still, extending its prerogatives and warnings into law and religious canon and into the minds of those who would abide by them. Because of its implications for survival and reproduction, humans find reputations immensely important, particularly as social animals whose survival is highly dependent on their relationships to others.

MEN

If humans are indeed the most dangerous animals, as philosopher David Livingston Smith suggests,[7] then it is thanks to the behaviors of men. With intellect and weapons, men have taken violent competition to the most stunning extremes. The history of warfare reveals eons of killing, torturing, and psychologically terrorizing rivals in huge numbers, utilizing means so creatively horrific as to truly boggle the mind. If we ever wish to understand these horrors of war, our scrutiny must be brought to bear on the psychology of male reputation (variously represented as pride, honor, face, dignity, respect, reverence, veneration, etc.). Following their primate ancestors, the reputations of men often relate to fitness concerns such as rank, fighting ability, territory controlled, and mate competition.

Like chimpanzees, men give nonverbal displays to show dominance. Men in bars, for example, may wear muscle shirts, talk loudly, and make large gestures, all in an effort to send dominance signals to other men. But language opens up wider channels of conveying status. Perhaps the most unambiguous examples are found in the titles of dominant men. The names of powerful male conquerors who reigned across world history convey extensive information about dominance rank and fighting ability. Alexander *the Great* (king of Macedon), for instance, was given this name not because he was a great dancer but a great conqueror. Many other designations leave less to the imagination: Ivan *the Terrible* (czar of Russia), Ismail *the Bloodthirsty* (of Morocco), Vlad *the Impaler* (of Romania), Alfonso *the Battler* (King Alfonso I of Aragon), or Nicolas *the Bloody* (Czar Nicholas II of Russia). Many dominant male leaders have earned names like *the Brave*. Alfonso III and IV of Portugal, Selim I of the Ottoman Empire; or *the Conqueror*. Alfonso I of Portugal, James I of Aragon, William I of England, and the entire Spanish military in the campaigns of the "New World." Names like these reflect violent reputations, and they perpetuate such reputations across time.

The titles of dominant men may also convey territorial dominion. The Romans started the precedent of awarding names (known as a *victory titles*) based on the territories a particular dominant male conquered. For instance the title of Scipio *Africanus* was awarded to the Roman general Scipio after he defeated Hannibal to gain control over Carthage. Other Roman victory titles include *Numidicus* ("the Numidian"), *Isauricus* ("the Isaurian"), *Creticus* ("the Cretan"), *Gothicus* ("the Goth"), *Germanicus* ("the German") and *Parthicus* ("the Parthian"). Victory titles can also convey the domination of people within a territory; for example, the title awarded to Charlemagne, the king of the Franks, was *Dominator Saxonorum* (Dominator of the Saxons).

Importantly, earning violent reputations conveys fitness benefits. As one example, Ismail the Bloodthirsty is reported to have kept 500 concubines and sired 888 children. Now, because reputations have such great implications for survival and because natural selection has engineered males to engage in violent mate competition, imagine the penalty for walking up to Ismail and calling him a "pussy" in front of his subjects. His reputed thirst for blood would be quenched from the neck of your decapitated body for everyone to witness. Slights to a man's reputation for dominance tend to bring about violence, if, that is, the particular male in question has the *power* to maintain his reputation through violence.

The value of reputation is equally evident in contemporary men. In North Korea, defaming the image or reputation of the cult leader Kim Jong Il has resulted in imprisonment or execution. Interestingly tied to inclusive fitness, in the North Korean system, punishment extends down genetic lines, meaning that children of prisoners inherit the prisoner status of their parents.[8] Similarly, if someone slanders the king of Jordan, they can be imprisoned and forced to endure three years of hard labor. Recent arrests have been made for this offense.[9] In Laura Betzig's study sample, Samoans who disrespected the king were tied up in front of an oven—symbolizing that the offenders were about to be roasted like pigs—and forced to eat human feces, ostensibly a habit of pigs (infrahumanization). These punishments were documented in court ledgers as recently as 1981.[10] Regarding the reputations of everyday men, Canadian evolutionary psychologists Martin Daly and Margo Wilson have explained:

> A seemingly minor affront is not merely a "stimulus" to action, isolated in time and space. It must be understood within a larger social context of reputations, face, social status, and enduring relationships. Men are known by their fellows as "the sort who can be pushed around" and "the sort who won't take any shit," as people whose word means action and people who are full of hot air, as guys whose girlfriends you can chat up with impunity or guys you don't want to mess with. In most social milieus, a man's reputation depends in part upon the maintenance of a credible threat to violence.[11]

Revenge is one way that threats are kept credible. In many cultures, men are expected to seek revenge if wronged or threatened, and revenge itself is awarded with honor, prestige, and power. On the other hand, *not* seeking revenge may be viewed as disdainful inaction, something that creates shame, disgrace, and a reputation for weakness. And the consequences are not only in the judgments of others; having a reputation for weakness can serve as an invitation to attack—a rule that men and other male primates understand well. Because these punishments exist for *not* seeking revenge, males are often trapped in endless blood feuds, where one act of retaliation is recompensed with another. These phenomena have been studied extensively, particularly in so-called *honor cultures*, cultures that place considerable emphasis on a man's honor. Historically, the American South supported a culture in which duels were fought to the death by men in the defense of honor. Research finds that southern men to this day continue to endorse greater eagerness to use violence to redress injuries to honor than do control subjects from other regions.[12]

But the importance of reputation is not only restricted to a geographic region—it is notably tied to male status competition nearly everywhere men are found, from the very bottom to the very top of the socioeconomic strata. Steven Pinker points out that even famous American historical figures have engaged in deadly competitions for the sake of honor. Notably, Treasury Secretary Alexander Hamilton was killed in a pistol duel by Vice President Aaron Burr, and President Andrew Jackson won two such duels and attempted to start several others.[13]

There is also a rich literature on the honor cultures of inner-city gangs. Like the dominant leaders of history whose honor status and territories are captured by victory titles, gangs often bear the names of their gangland. Here sociologist Andrew Papachristos relays the words of "Melo," one mid-ranking leader of the urban *Two Six Nation* gang of Chicago—"Two Six" refers to 26th Street, where the gang's territory originated:

> This is *our* 'hood, see? We got no choice but to protect it. If we back down, we ain't shit. Everyone will think we ain't nothin' but a bunch of punk ass bitches. . . . How can we call ourselves 2-6, if we don't got this corner? We always had this spot. . . . Without that, what do we got? Nothing. Might as well join the fucking Boy Scouts if you ain't got a spot.[14]

Melo's reference underscores an important quality of group affiliation—that groups operating on dominant-male ideologies assume rank among other groups in the manner of individual men among other men. Papachristos points out that as a result, "disputes often become intrinsically collective because the group regards an offense against a member as an offense against all, a sentiment that fosters in-group cohesion as a function of confronting external threats."[15]

Writ large, what this essentially means is that we have gangs, tribes, states, and even nations operating on the geopolitical scene like individual men bent on protecting their reputation as badasses. Ideologies such as "national honor" easily arise from these male strategies, and behaviors within the global society of nations take on the same character as two men squaring off for violent competition. Nations, like individuals, may also become trapped by the potential consequences of backing down. German historian and political writer Heinrich von Treitschke (1843–1896) made this observation of nations using the idiom of male psychology: "When the name of the State is insulted, it is the duty of the State to demand satisfaction." Without an immediate apology, which is one way to restore honor, war must follow, "no matter how trivial the occasion may appear, for the State must strain every nerve to preserve for itself the respect which it enjoys in the state system."[16]

Von Treitschke's observations are recapitulated in the geopolitical arena. Former Secretary of State Henry Kissinger, for example, wrote in his memoirs, "No serious policymaker could allow himself to succumb to the fashionable debunking of 'prestige' or 'honor' or 'credibility,'"[17] suggesting that honor was an important motivation for keeping the United States in the Vietnam War. Indeed, one week before sending hundreds of thousands of American troops to Vietnam, President Lyndon B. Johnson publicly announced, "Our national honor is at stake in Southeast Asia, and we are going to protect it."[18] The role of honor continues to be a focal concern in the politics of war-making. For example, Deputy Defense Secretary Paul Wolfowitz pondered of the Iraq war, "I think that already . . . the magnitude of the crimes of that regime and those images of people pulling down a statue and celebrating the arrival of American troops is having a shaming effect throughout the region."[19]

Recall that shame (as opposed to pride) has important implications for rank status. Insults to honor (which create shame) can incite violence, and military strategists value cultural knowledge for the map of potential social landmines it provides. But it should also be understood that although insults to male honor can theoretically take many forms, they almost invariably flow through existing Darwinian channels. For instance, historian Bertram Wyatt-Brown points out that the video made of Saddam Hussein after his capture showing his beard being checked for lice was a deliberate attempt to show him dishonored in a culture where even lightly touching a man's beard is considered a severe insult (something we have discussed).[20] Surely the video incensed the Ba'athist, Sunni, and foreign Islamists who saw it and incited them to retaliate with violence. Further, Wyatt-Brown argues that when US Ambassador Paul Bremer dismantled the Iraqi Army and the Ba'athist bureaucracy, these actions had the effect of dishonoring the men of Iraq by removing their ability to protect their women:

> On another level, Bremer had not only humiliated considerable numbers of men but also denied them the ability to shield their women from the possibility and infamy of assault and rape. Protec-

tion of women's honor, *ird,* inflames Iraqi males to near obsession. That is because in Middle Eastern cultures women are judged the very center of male ownership rights. Whether true or not, rumors that American soldiers take Iraqi women into their tanks and Humvees for lovemaking or rapes are pervasive in Baghdad. Whoever the rapist may be, to dishonor the woman in that fashion is to disgrace her and her kindred. In much of the region, to restore family honor, relatives feel required to kill the victim of rape, no matter what extenuating circumstances there might be.[21]

Wyatt-Brown further discusses how, two months before the 9/11 attacks on the World Trade Center, Arab viewers faxed the Qatari television host al-Qassem the following opinion of Osama bin Laden: "In light of the terrible Arab surrender and self-abasement to America and Israel, many of the Arabs unite around this man, who pacifies their rage and restores some of their trampled honor, their lost political, economic, and cultural honor."[22] Wyatt-Brown also claims that when the United States withdrew from Islamic Somalia, it gave courage to terrorists (recall here the strategic risk of backing down). About this, bin Laden declared, "We found [the Americans] had no power worthy of mention. . . . America exited dragging its tails [*sic*] in failure, defeat, and ruin, caring for nothing."[23]

President George W. Bush (who claimed that God told him to invade Iraq) called the Iraqi campaign a "crusade." Recalling the bloodshed, rape, and tyranny of the actual Crusades, this was a slap in the face to many in the Islamic world. Bin Laden responded in turn by attacking honor (italic mine): "If the Americans refuse to listen to our advice . . . then be aware that you will lose this Crusade Bush began, just like the other previous Crusades in which you were *humiliated* by the hands of the Mujahideen, fleeing to your home in great silence and *disgrace.*"[24]

With minimal effort, we could extract a litany of similar allusions to shame and honor from the rhetoric of nations. However, it might better benefit our understanding to listen for the expressions of honor as reiterated across the canons of religion. Just as men recognize and value "national honor" and are willing to fight to defend it, men will also fight for the masculinized honor of their religions.

GODS

God is portrayed as having concern for his fearsome reputation and that concern follows the pattern of male primates. Violence committed in defense of God's reputation (variously expressed as his name, honor, respect, reverence, veneration, etc.) often follows primate patterns of revenge, sexual control, and rank-maintenance—although an omnipotent, everlasting being should logically show no personal interest in any of these fitness-related concerns. Further, just as powerful men have dictated that any criticism against them shall be punished, religions have created proscriptions against criticizing God, and such rules have been enforced with savage violence.

One means of ensuring the legitimacy of God's position of power has been to forbid the questioning of religious doctrines that support it. Because our brains evolved in social environments that necessitated acquiescing to dominant males, often the fear and respect that motivate *not* questioning are experienced so automatically that understanding why requires making the natural seem strange. For example, Richard Dawkins has argued:

> A widespread assumption, which nearly everybody in our society accepts—the non-religious included—is that religious faith is especially vulnerable to offence and should be protected by an abnormally thick wall of respect, in a different class from the respect that any human being should pay to any other.[25]

This is a brilliant point. Often respect for religion means never questioning its tenets, but there are good reasons to subject religion to the same scrutiny as any other human enterprise. Given the propensity of religions to engender violence, not only is there no rational reason why we shouldn't discuss religion, but such discussions are our moral obligation. I would add, however, that there is indeed a class of respect similar to that paid religion, and that is the respect paid to powerful men, around whom there are Dawkins's "abnormally thick walls." Apart from the fact that exploring religion threatens to expose certain unpleasant existential realities, it also touches upon our evolved psychology in which fear of and respect for powerful males are deeply embedded. And fear, respect, and caution have much to do with upholding the reputation of God, much as they do the reputations of men.

Like men, God shows concern that his reputation for violence is spread across his land, and he builds his reputation for greatness in the manner of dominant men such as Peter the Great, or Alexander the Great, or any of the others—through killing:

> For I will at this time send all my plagues upon thine heart, and upon thy servants, and upon thy people; that thou mayest know that there is none like me in all the earth. For now I will stretch out my hand, that I may smite thee and thy people with pestilence; and thou shalt be cut off from the earth. And in very deed for this cause have I raised thee up, for to shew in thee my power; and that my name may be declared throughout all the earth. (Exod. 9:14–16)

Like men, God is exceedingly intolerant of threats to his reputation, most seminally in the third commandment: "Thou shalt not take the name of the LORD thy God in vain; for the LORD will not hold him guiltless that taketh His name in vain." (Exod. 20:7). Words spoken against God (or disrespecting, slighting, or otherwise not showing proper reverence) are thereafter forbidden and considered a crime known as *blasphemy*. This crime is also punishable by death. One biblical passage recounts a fight between two men in which one of them blasphemes, "And the Israelitish woman's son blasphemed the name of the Lord, and cursed" (Lev. 24:11). The blasphemer was brought to Moses, whose task it was to reveal to the people what God wished them to do about the transgression. God commanded Moses to have the people of the camp bash the blasphemer's head in with stones:

> The Lord spake unto Moses, saying bring for him that hath cursed without the camp; and let all that heard him lay their hands upon his head, and let all the congregation stone him . . . he that blasphemeth the name of the Lord, he shall surely be put to death. (Lev. 24:13–16).

For the sake of reputation, the God of the Old Testament has also been known to kill children. Here, his concern over reputation is richly embedded within several other evolutionary scripts. As the story goes,

David killed Uriah in order to win sexual access to Uriah's wife. This was a greedy move of which God, being a dominant male, did not approve. In response he kills the child David had with Uriah's widow. It is noteworthy that God didn't seem to have a problem with David killing Uriah so much as David taking Uriah's wife and reproducing with her. Dominant males often work to contain the sexual ambitions of their subordinates. It is telling that, when the prophet Nathan goes to tell David what God will do to him for threatening his reputation, he says:

> [B]ecause by this deed you have given occasion to the enemies of the LORD to blaspheme, the child also that is born to you shall surely die. So Nathan went to his house. Then the LORD struck the child that Uriah's widow bore to David, so that he was very sick. David therefore inquired of God for the child; and David fasted and went and lay all night on the ground. . . . Then it happened on the seventh day that the child died. (2 Sam. 12:14–18)

Notably, David's first impulse was to demonstrate subordination by surrendering food resources to God. In the end, God kills David's child with sickness, thus punishing David for reproducing with his stolen female. This shows what we have discussed about how dominant males will commit infanticide against the offspring of rival males. But also of great concern is the fact that God took such a brutal step lest David's actions cause his name to be blasphemed by his enemies. Apes and men both turn aggressive when aid is given to rivals, and men become violent when aid takes the form of defamation. Here God does as well and in effect kills a child to protect his fearsome reputation. All this killing boils down to the fact that this concept of blasphemy, in the end, reflects that enduring male primate concern over dominance reputation, projected onto a dominant male god.

Christ was also mindful of male reputation. For example, he is purported to have returned sight to a blind man, after which the Pharisees began to challenge his reputation, claiming that his healing was conducted using the power of Beelzebub, the ruler of demons. Christ responded with a threat in turn:

> Therefore I say to you, any sin and blasphemy shall be forgiven men, but blasphemy against the Spirit shall not be forgiven. And whoever

shall speak a word against the Son of Man, it shall be forgiven him; but whoever shall speak against the Holy Spirit, it shall not be forgiven him, either in this age, or in the age to come (Matt.12:31–32).

The Pharisees were challenging the very legitimacy of Christ's ministry by accusing his work of being the design of Satan, a charge that could *not* be forgiven. Given the future described in the book of Revelations, one can only suppose that "the age to come" is a world ablaze with retaliatory fury. Whether or not Christ (or legends of Christ) meant these things will be forever left to speculation. What is certain is that the man portrayed as giving these admonishments knew the importance of reputation and was ready to defend it.

Among dominant kings and their servants there is disproportionate value placed on the honor of those possessing high rank— thus Kim Jong Il was *not* executed for slander, the Samoan kings were *not* required to eat feces, etc. This disparity is also conspicuously described in religious doctrine. For instance, inspired by the passage above from Matthew, Thomas Aquinas took the liberty of projecting the concerns of man onto God by declaring that a verbal slight against God is a more grievous crime than murdering another human. He argues, "It is clear that blasphemy, which is a sin committed directly against God, is graver than murder, which is a sin against one's neighbor."[26] God's immaterial reputation takes on a value greater than human life.

Although the notion of divine retribution has been a reliable Christian fallback over time, with Judgment Day an ever-present threat to those who might ignore the demands of faith, Christians of certain eras did not abstain from taking punishment for blasphemy into their own hands. The Spanish Inquisition was infamous for torturing blasphemers with an expansive toolset of sadistic devices. It was not unheard of to cut out the tongue of blasphemers or to pierce the tongue with a steel rod. Often blasphemers were exiled or jailed. Scourging, a practice of publicly beating blasphemers with a multi-thong whip, was common. Some Jews who dared to blaspheme were forced to wear bridles,[27] something akin to the punishment of the Samoans, except that blaspheming Jews were made figurative beasts of burden rather than pigs. A long history of execution for blasphemy is also recorded across Christendom. Britain's last such execution took place in 1697, when twenty-year-old Thomas

Aikenhead was killed for denying the truth of the Old Testament and questioning Christ's miracles.[28]

Once again, the manner in which we view God as experiencing pride, craving respect, or being jealous of respect paid to others rests on our evolved psychology and is based on competitions among men. This may explain why religions have created rival figureheads to represent rivaling ideologies (e.g., God and Satan) —so that people can understand abstract religious ideals using psychology already designed to understand rivalling men. Such ideals without archetypal figures to represent them are difficult to understand and do not evoke the same emotional resonance—amorphous concepts require being congealed into human form. Yahweh, like a headman of a Judaic tribe, was known to have numerous male rivals, such as Baal, Chemosh, Astarte, Milcom, and many of the gods of Egypt. In establishing God's supremacy some have claimed that these figures are only "empty idols," rather than gods. However they are defined clearly as other gods throughout the Bible, and as gods they are rivals, as threatening to God, and possibly also God's people, as the endless succession of male threats across human evolutionary history. Yahweh therefore defends his reputation against these archetypal enemies, and in the manner of men, he goes to great lengths to assure no other gods receive respect, honor, worship, or recognition from his subordinates.

Jews and Christians are not alone in creating honor cultures for God. With Islam's Judaic origins, it follows that we also witness feverish efforts to protect God's reputation in its religious doctrine. However, in the Koran there is no specific prescription for humans to punish blasphemers. Allah is described as punishing blasphemy himself:

> But indeed they uttered blasphemy. . . . If they repent, it will be best for them; but if they turn back (to their evil ways), Allah will punish them with a grievous penalty in this life and in the Hereafter. (Koran 9:74)

It is important to note, however, that tolerance of dissenting perspectives is actually advised in the Koran in a manner that contrasts sharply with the reactionary behaviors of some of the more extremist Muslims of the world today:

Bear, then, with patience, all that they say, and celebrate the praises of thy Lord, before the rising of the sun and before (its) setting. (Koran 50:39)

Sharia law—a set of precepts inspired by the Koran (about which there are many diverging interpretations across Islam)—is another story; it awards Muslims the authority to flog, amputate, behead, or hang blasphemers. Armed with Sharia law, certain Islamist zealots go about acting as though their God's reputation requires as fearful of a defense as that of a mortal man vulnerable to physical attack. They defend Allah's reputation by killing those they interpret as disrespecting him. In the year this book was written, Afghanistan, Pakistan, Saudi Arabia, Yemen, and other countries continue to allow the killing of blasphemers, some under Sharia law and some, like Pakistan, following actual penal codes based on Sharia law. The manner in which blasphemy is interpreted in modern Islam has come to resemble the witch hunts of Christian medieval Europe. There are numerous examples that astound the rational mind.

In 2007, for example, a British schoolteacher in Sudan, Gillian Gibbons, was imprisoned for the infamous teddy bear blasphemy case. Her crime was allegedly to allow her class to name a teddy bear "Muhammad," which incensed Islamic zealots in Sudan. A hoard of angry protestors ten thousand strong flooded the streets of Khartoum, many of them brandishing swords and machetes, demanding Gibbon's execution.[29] Lucky for her, the protestors didn't have their way and Gibbons got away with a brief jail sentence before getting deported. Unfortunately, there have been countless lives taken throughout the Islamic world for allegedly insulting God. Many of the charges turn out to be false—for example, it was later discovered that the teddy was actually named after a student in Gibbon's class rather than the prophet Muhammad.

Often, rampaging mobs indeed get their way. When the zealot Florida pastor Terry Jones taunted Muslims by burning a copy of the Koran, a mob attacked a United Nations compound in Mazar-e-Sharif, Afghanistan, killing seven UN employees. Violent protests followed in Kandahar, leaving nine dead and over eighty others injured.[30] The pointless murder of the UN employees who had nothing to do with this largely insignificant Florida man illustrates the darkly irrational side of religious belief. Further, like the God described by Thomas Aquinas, Allah's reputation is also measured in human lives.

In February 2012, several copies of the Koran were accidentally incinerated by NATO officials at Bagram Air Base in Afghanistan. Incensed Muslims soon began rioting, burning, and killing, and in the end, thirty Afghanis and four American soldiers lay dead.[31] One protestor outside the gate of the air base told reporters, "They have burned our Holy Korans. . . . This means they burned our faith, our honor and our lives."[32] One month later a US Soldier wandered off base into a village in Kandahar and murdered sixteen innocent civilians in cold blood, including nine children. By comparison, the national response to this senseless mass killing was tepid at best. Mullah Khaliq Dad, a member of the council in charge of the investigation, was asked why Afghan streets were not ablaze with fires and revenge killings as they were when the Korans were burnt one month prior. Seemingly incredulous he responded, "How can you compare the dishonoring of the Holy Koran with the martyrdom of innocent civilians? The whole goal of our life is religion."[33] Reactions to the name of a dominant male (as in the teddy bear incident) and his edicts (as in the Koran) strongly suggest religions are honor cultures, much like in the streets of Chicago, the American South, or the rainforests of Gombe.

In part because honor is tightly interwoven with religious sentiment, fighting jihad has a strong pull for many Muslim youth. Indeed, many Islamist leaders teach that jihad is the only way to achieve honor and dignity. There have been obvious parallels in Christian history, times when killing or dying in the name of Christ brought high honor, particularly in the age of the Crusades. But modern-day Islam stands apart from the other Abrahamic faiths for remaining so apparently beholden to medieval ideologies of this kind. With honor killings snuffing out the lives of women for the sake of their offended husbands, brothers, or fathers, with raging mobs willing to kill over stuffed animals, and with suicide bombers praised for returning honor to their Muslim survivors, we are left to conclude that male dignity among many Muslims is exquisitely fragile, as is the projected male dignity of Islam.

But extremist sentiment also emerges from contemporary Christian mouths, many of whom also appear to give voice to the most primitive of urges. The American attorney and political commentator Ann Coulter infamously wrote in response to 9/11:

We should invade their countries, kill their leaders and convert them to Christianity. We weren't punctilious about locating and punishing only Hitler and his top officers. We carpet-bombed German cities; we killed civilians. That's war. And this is war.[34]

Ann Coulter's response mirrors that of many other Christian thinkers, and illustrates that at times women can adopt male-typical strategies of revenge—after all, such strategies work for female macaques. The only problem is that, in the long term, reputational arms races tend to escalate rapidly and end slowly. As one example, the contest over honor has kept Islam and Christianity at odds with one another for centuries, and has written a long history of despotic behaviors between them.

It would be an oversimplification, however, to ignore the fact that beyond needing violent reputations, dominant males are also required to build reputations for trustworthiness and cooperation if they are to retain power over time. This is seen among nonhuman primates—alpha male chimpanzees, for instance, may be punished en masse by their subordinates for using too much violence.[35] Even despotic men may develop reputations for cooperation, and in addition to earning names like *the Bloodthirsty* and *the Impaler*, men have also won names like *the Fair* (e.g., Charles IV of France, Donald III of Scotland, Ivan II of Russia) or *the Just* (e.g., James II of Aragon, Louis VIII of France). Whether or not these men deserved their titles is another matter.

We seem to have a similar investment in seeing God as a trustworthy cooperator. God is described as having the qualities humans most desire in an ally:

> The Rock, his work is perfect, for all his ways are justice. A God of faithfulness and without iniquity, just and upright is he. (Deut. 32:4)

Islam takes a similar approach. God is described in the Koran as "oft-forgiving, most merciful, most gracious" innumerable times, and often these descriptors accompany his spoken name. Viewing God as just, fair, and righteous is a perspective taken across religious traditions that speaks to the rules of reciprocal altruism projected onto God.

Nevertheless, the pattern of violence surrounding God's reputation persists, and its human designs must be delineated. Not only are man's rules projected onto God, but powerful men have secured power by conflating God's reported concern with reputation with their own. For instance, "You shall not revile God, nor curse a ruler of your people" (Exod. 22:28). Men have engineered these conflations to address Darwinian imperatives such as dominance rank:

> Let as many servants as are under the yoke count their own masters worthy of all honour, that the name of God and his doctrine be not blasphemed. (1 Tim. 6:1)

And sexual control:

> Teach the young women to be . . . obedient to their own husbands, that the word of God be not blasphemed. (Titus 2:4–5)

And lest it be forgotten, all the killings committed to defend God's name were conducted by men in the manner of men defending their own reputations. One can be sure that such killings reinforced the power and rank positions of men, along with their own fearsome reputations, and secured for them other evolutionary interests.

Often the religious and nonreligious alike are shocked at the atrocities committed in defense of God's name, and the shock is riddled with befuddlement at how capricious, senseless, and anachronistic such acts of violence are in the modern age. Through evolutionary science we can gain an understanding of God's overzealous concern with reputation and how it is based on an ancient legacy of primate-male mate competition. We can also begin to grasp the minds of men who are ready to kill to defend his honor and who proscribe blasphemy upon pain of death. In order to curb religious violence committed in defense of God's name, we must not grovel to male dominance, but question its place in our religions and thoughtfully dictate its boundaries.

Chapter 9

GOD'S TERRITORY

"It's about our D and A. Descendants and ancestors. We are the descendants and we are the ancestors. D and A, our DNA, our blood, our flesh and our bone, is made up of the metals and the minerals and the liquids of the earth. We are the earth. We truly, literally and figuratively are the earth." —John Trudell, Native American activist

Science shows us that we are indeed made of the earth, of its minerals and liquids and metals. But life also requires useable energy. Energy is what makes earth material able to organize and perform the functions required for life. Energy is the very thing that sustains organizational integrity, keeping organic beings from dissolving back into the strata while living. All life forms derive this energy from the sun. Humans draw upon sun energy captured in plants and other animals—this makes humans truly sun energy as well as truly earth.

Because useable sun energy on earth is finite, life competes for it. The notion of territory is a kind of quale that energy-requiring beings project onto geographic space, signifying the right to utilize energy resources within specified boundaries. This right is won through competition. In this chapter we will therefore discuss territory not only in terms of tracts of earth but also in terms of competition for the vital resources they contain. And yet again, if we take scripture at face value, we face an indeterminable contradiction—here, the idea of the Abrahamic god, who exists in an ethereal plane and requires neither the earth nor the earth's yield to survive, showing such zealous interest in territory. The ultimate answer for this bestial concern can once again be traced back to our own evolutionary history.

MARKING TERRITORY

To understand the innate, animalistic quality of God's territoriality, we must revisit its origins. We begin with the matter of marking territory, a behavior that a vast number of species engage in using sound (e.g., bird song, monkey vocalizations), visual threat displays, or scent (by way of urine, droppings, or specialized scent glands). Males (and sometimes females) may patrol their demarcated territories and will defend them by force if necessary. Often, however, territorial displays or markers allow individuals to avert costly violence; they serve as warnings for competitors to stay away, and thus avoid potentially damaging physical conflict. When these markers are ignored, the invading individual is typically willing to risk violent confrontation for the potential gain of appropriating some desired resource: food, water, or females. When conflict does arise and the challenger is victorious, he will mark the territory as his, covering all traces of his predecessor.

Human beings are intensely territorial. Today there are 192 countries in the world, all with delineated borders, all patrolled and defended by armed men (and the occasional woman). Countries are distinguished by national flags, which serve as visual cues conveying information about dominance, territorial control, and in-group loyalty. The American flag, for example, is suffused with symbolic meaning of this sort: the thirteen red and white stripes symbolize the thirteen colonies that rebelled against their dominator, the Kingdom of Great Britain, to gain control of its territories; the fifty stars represent the collective territories that now share in-group commitment, particularly in alliance against outsiders. By way of symbolism, the flag says in essence: *This is how we won it—and this is the loyalty we share against any usurpers.* As such, flags often serve to warn outsiders. Further, flags of all nations across the ages have had a central role in military campaigns and are used to mark territory acquired in war. Last, flags are regarded as sacrosanct, and their desecration provokes outrage and sometimes violence, much in the manner of religious symbols.

Not surprisingly, religious symbols often serve the same function as flags or even earlier forms of marking—to communicate dominance over territory, including that which is acquired through conflict. During the bloody Spanish conquests of the Americas, Spanish forces staked crosses all throughout the New World, rather explicitly claiming territory for the Catholic Church. Upon capturing a city, the Conquistadors went about methodically smashing idols, dismantling altars, and leveling the religious temples of the indigenous peoples they dominated. They planted crosses and erected cathedrals atop the ruins, flooding their claims with Catholic iconography—all to provide the conquered nations a clear sign of the new male god's dominance. This took place in virtually every indigenous community with which they had contact, from the tip of South America northward to California, Arizona, New Mexico, and Texas. A good example is the Metropolitan Cathedral of Mexico City, the largest and oldest cathedral of the Americas. This towering, gilded, stone temple was built to demonstrate the superiority of the Christian god over the resident gods of the Aztec people after the Spanish conquered the capital city of their civilization, Tenochtitlán. To emphasize the point, the cathedral was built atop the demolished *Templo Mayor* (Major Temple) of the Aztec—a temple dedicated to the male gods of war and rain—using the former temple's stones. The Spanish often boasted about the power of their god to the dignitaries of the civilizations they subjugated and warned the Indians to fear, submit to, and worship their god in the place of all other religious figures. Crosses were described to the natives as the superior (male) god's marker.[1]

The practice of destroying and effectively marking over the resident gods' markers was not new to the conquistadors. Yahweh, for example, is reported to have commanded Judaic tribes to obliterate any accoutrements of conquered religions: "Break down their altars, smash their sacred stones and cut down their Asherah poles" (Exod. 34:13).

Islam is no exception, and members of the more radical factions have followed the same patterns, albeit in more recent history. The Buddhas of Bamiyan in central Afghanistan were the largest standing statues of the Buddha in the world. They were carved into the side of a mountain in the sixth century in Bamiyan, an ancient city along the Silk Road that was once a thriving Buddhist religious center. When the territory was conquered during the Islamist invasions of the seventh century, the statues were left intact by the invaders. However, when over a millennium later the Taliban came to power exhorting Sharia law, a set of religious precepts espousing particularly exaggerated forms of male dominance, these giant (male) figures were deemed a threat and subsequently destroyed with rockets and dynamite, despite outcry from around the world. The leader of the Taliban Mullah Mohammed Omar was quoted as stating, "Muslims should be proud of smashing idols. It has given praise to God that we have destroyed them."[2]

In an evolutionary context, it is consistent that a religion based on a dominant male god and represented by a dominant male prophet should seek to destroy the effigy of another deified male. Similarly, in July 2012 factional warfare in Mali resulted in victory for the Islamist Movement for Oneness and Jihad in West Africa (MOJWA), along with its ally Ansar Dine (translated, *Defenders of Faith*). Shortly after their victory, the Ansar Dine began taking sledge hammers and pickaxes to seven ancient World Heritage site mausolea in Timbuktu, some that had been standing since the fourteenth century. The shrines were dedicated to local (male) saints and deceased (male) sages which the militants reportedly took issue with, declaring them "haram" or *displeasing to Allah*. The shrines and the cultural treasures inside were smashed to pieces.[3]

While religious symbols are erected and dismantled to demonstrate territorial dominance much in the manner of national flags, they appear to take the conveyance of power a step further by exaggerating male dominance *through God* as a means to dissuade would-be invaders. In other words, dominant men may be fearsome, but dominant gods possess the power to smash dominant men like ants; no small threat in the minds of believers. Because humans descend from ancestors for whom alliances with powerful males served as an effective deterrent to potential aggressors, claiming alliance with a being of great power is intuitively meaningful, particularly when that being is portrayed as a dominant man operating by familiar evolutionary rules.

Territory is central to the narrative of Genesis, the foundational credo of the three great Abrahamic faiths. Stripped down, the narrative recapitulates ancient patterns of male-typical territoriality. In the story we have a dominant male (God) ruling over a resource-laden territory (the Garden of Eden). In this world another lesser male (Adam) emerges and defies the rules of the dominant male's (God's) territory by taking food (fruit from the tree of life). As we have discussed, different interpretations of that forbidden fruit exist, but all have evolutionary significance. The fruit has been described variously as food, as sex, as knowledge, and as the presumption to God's position of power— "For God knows that when you eat from it your eyes will be opened, and you will be like God" (Gen. 3:5).

Suffice it to say, dominant males generally prefer to have control over all such privileges. For his transgressions, the lesser male (Adam) is punished with pain and death—God making Adam mortal is tantamount to killing him. Conceivably the story of Genesis could involve a narrative wholly foreign to animal notions of male dominance, but remove the parentheticals above and the story, once again, could just have easily occurred among chimpanzees in the forests of Gombe.

Another theme central to Christianity and Islam is the rivalry between God and Satan over their respective territories. As the story goes, Lucifer, like Adam, was banished from God's territory for aspiring to the ascendancy of his throne, and now resides in a less bountiful territory, the *depths of the pit*, or hell. Like any good rival, however, Lucifer is said to lie in wait, anticipating the moment when he can rise to challenge God once more.

The god described as having concern for his own territories also has concern for the territories of the dominant men who represent him. The Old Testament does not lack for references to territory. In the Mosaic covenant described in Exodus, a dominant male (God) comes to an agreement with his subordinates (the Israelites) that they will obey his commandment to not philander with rival males (e.g., the gods of Canaan). In return the dominant male (God) agrees to win territory for his subordinates, displace resident competitors, and protect that land from potential rivals:

> Obey what I command you today. I will drive out before you the Amorites, Canaanites, Hittites, Perizzites, Hivites and Jebusites. . . . I will drive out nations before you and enlarge your territory, and no one will covet your land when you go up three times each year to appear before the LORD your God. (Exod. 34:11, 24)

Similarly, after Moses's death, God placed Moses's son Joshua in power and through him granted vast territories to early Judaic tribes and protected those territories from rivals:

> I will give you every place where you set your foot, as I promised Moses. Your territory will extend from the desert to Lebanon, and from the great river, the Euphrates —all the Hittite country—to the Mediter-

ranean Sea in the west. No one will be able to stand against you all the days of your life. As I was with Moses, so I will be with you; I will never leave you nor forsake you. Be strong and courageous, because you will lead these people to inherit the land I swore to their ancestors to give them. (Josh. 1:3–6)

God continued the allocation of land to Joshua in his old age, supporting Joshua in conquest as he deposed dominant male rulers and appropriated their lands, defining his new territorial boundaries with great precision:

> When Joshua had grown old, the LORD said to him, "You are now very old, and there are still very large areas of land to be taken over. This is the land that remains: all the regions of the Philistines and Geshurites, from the Shihor River on the east of Egypt to the territory of Ekron on the north, all of it counted as Canaanite though held by the five Philistine rulers in Gaza, Ashdod, Ashkelon, Gath and Ekron; the territory of the Avvites on the south; all the land of the Canaanites, from Arah of the Sidonians as far as Aphek and the border of the Amorites; the area of Byblos; and all Lebanon to the east, from Baal Gad below mount Hermon to Lebo Hamath. As for all the inhabitants of the mountain regions from Lebanon to Misrephoth Maim, that is, all the Sidonians, I myself will drive them out before the Israelites. Be sure to allocate this land to Israel for an inheritance, as I have instructed you, and divide it as an inheritance among the nine tribes and half of the tribe of Manasseh." (Josh. 13:1–7)

In each of these examples Yahweh, a dominant male God, references a sequence of evolutionary scripts—promising to secure territory for his subordinates in exchange for obeisance, subservience, and loyalty. Here he promises a lush and bountiful territory, provisioned with rain from the heavens—but he warns the Israelites not to make submissive displays to other gods, or he'll kill them by causing drought and starvation:

> Observe therefore all the commands I am giving you today, so that you may have the strength to go in and take over the land that you are crossing the Jordan to possess, and so that you may live long in the land the LORD swore to your ancestors to give to them and their descen-

dants, a land flowing with milk and honey. The land you are entering to take over is not like the land of Egypt, from which you have come, where you planted your seed and irrigated it by foot as in a vegetable garden. But the land you are crossing the Jordan to take possession of is a land of mountains and valleys that drinks rain from heaven. It is a land the LORD your God cares for; the eyes of the LORD your God are continually on it from the beginning of the year to its end.

So if you faithfully obey the commands I am giving you today—to love the LORD your God and to serve him with all your heart and with all your soul — then I will send rain on your land in its season, both autumn and spring rains, so that you may gather in your grain, new wine and olive oil. I will provide grass in the fields for your cattle, and you will eat and be satisfied.

Be careful, or you will be enticed to turn away and worship other gods and bow down to them. Then the LORD's anger will burn against you, and he will shut up the heavens so that it will not rain and the ground will yield no produce, and you will soon perish from the good land the LORD is giving you. (Deut. 11:8–18).

To summarize my point thus far, humans have evolved in a social environment in which powerful males often led the way into new territories, helped secure the material resources therein, and provided protection over those territories from other male raiders. Historically these men have been backed by man-based gods. A Martian anthropologist with an outsider's perspective would study the territorial behaviors of monkeys, apes, and humans and find the same pattern of acquiring, granting, and protecting territory in the biblical god with no difficulty.

However, the power of God as a dominant ape goes beyond that of most other male primates, in that when he defines territories on the earthly plane, they become *holy lands*. For the Abrahamic faiths, this notion of holy land culminates in Jerusalem as in no other place in the world, and its holiness correlates to its number of territorial dissections. Contemporary Jerusalem is partitioned off by the three Abrahamic faiths, each with a shrine deemed intensely sacred. Muslims, for example, have the Dome of the Rock, the spot at which Muhammad is believed to have ascended into heaven. Christians have the Church of the Holy Sepulchre, where Jesus is said to have been buried. Jews stake claim to the Wailing Wall and the Foundation Stone—purportedly representing the very place where God began creating the entire universe. In addition, there are shared holy sites contested for by Jews, Christians, and Muslims—won and lost by each over the years—such as the Tomb of King David, which is now controlled by the Jews.

But like all human territories, the holy lands also have practical significance that betrays their human origins. Jerusalem is strategically located between three great land masses and serves as passage between Europe, Asia, and Africa. It is at the center of trade routes by land and sea through the Mediterranean, the Red Sea, the Persian Gulf, and the Indian Ocean. In the modern world, it is a central point of nations with rich petroleum and natural gas resources. Not surprisingly, for thousands of years control of the region has been the ambition of powerful kings and other (usually male) political figures willing to stake their claims by force of arms.

TERRITORY: STAKING CLAIM TO SEX

That human conflict over territory should concern energy resources (such as food, petroleum, etc.) stirs little controversy. From the small-scale raids of hunter-gatherers to nations that fight for trade routes, economic sanctions, or oil, the connection of warfare to resource competition is usually demonstrable even when obscured by national pride, ideology, religion, or other propaganda designed to motivate the masses to arms. But energy is only half the story. For beings fated to decline and die, programmed by nature to reproduce themselves into the future, conflict over territory also concerns sex. Sexual motivation for territorial violence is less obvious in humans than in nonhuman animals, and perceiving it generally requires, again, making the natural seem strange. To accomplish this, we shall start with the behaviors of nonhuman primates and work our way to men and their man-based gods.

In the rainforests of Sumatra and Borneo, orangutans live semisolitary lives with adult females living in slightly overlapping territories subsumed within the larger territory of a dominant male with whom the females preferentially mate.[4] Subadult males are often capable of reproduction but delay developing secondary sex characteristics such as face flanges in order to duck the violent attention of dominant males. These lesser males are usually transient, traveling outside the peripheries of larger, flanged, and more aggressive adult males' territories. Sometimes subadults will sneak into a dominant male's territory and force copulation on the females. Primatologists have argued that this behavior is tantamount to rape. While the designation *rape* has generated some controversy, primatologist Birutė Galdikas's field observations leave little question:

> Rape occurred when a male attempted to copulate or copulated with a female who resisted his efforts to position her for intromission. A female's struggle ranged in intensity and duration all the way from brief tussles with squalling and some pushing and slapping at the male's hand to protracted violent fights in which the female struggled through the length of copulation, emitting loud rape-grunts and bit the male whenever she could.[5]

Rape also occurs when a mature, flanged male invades another's territory when the resident male is out of sight. Apparently, territorial status is important to orangutan females, who usually (though not always) resist copulations with males from outside territories. Anthropologist John Mitani's research estimated that 90 percent of copulations with the resident, flanged males were consensual, whereas only 34 percent of the copulations with nonresident flanged males were—in other words, 66 percent were forced.[6] Generally, a mature male will make a long call with his laryngeal sacs as a warning for rival males to stay outside his borders. When his resonant bellows are ignored, competition, while not typically fatal, can be fierce and injurious, resulting in lacerations, missing eyes, and severed digits. When one flanged male overthrows another, he gains sexual access to the females within the displaced male's former territory.

Similarly, baboons vie for territory on the savannah, deserts, and steppes of Africa. The field of play is savage, and each competitor comes armed with five-inch canines (longer than a lion's). Rival troops of baboons fight for resources, such as preferred grazing land or watering holes. Chasing, biting, and bloodletting characterize the frays. Occasionally gangs of baboons will even target humans—attacking tourist caravans by opening car doors, jumping into windows, and stealing sandwiches and cookies, which the duly intimidated tourists usually surrender. So while it would not be fair to say that *all* competition has sex as its primary motivation, male baboons *do* battle for females, which they steal away from other males in combat. The dominant male adds stolen females to his harem, the members of which he strives to keep cloistered, fiercely battling any rival intent on stealing them. Other male primates also fight each other for sexual privilege, which is often conferred by controlling territory.

For a sophisticated level of organized warfare we look to chimpanzees, one species among few in the animal kingdom that engage in large-scale, organized coalitional violence—others include ants, dolphins, wolves, hyenas, lions, and humans. Every chimp knows when the war party is forming. The normally boisterous troop goes chillingly silent and tense, while visibly projecting a sense of shared intention. After a time the almost exclusively male squad sets out single file in the direction of the group borders, using a stereotyped gait reserved only for patrols. At the edge of their territorial boundary they start scanning the trees, looking out across valleys, listening for sounds of the enemy. When they find a lone male or a smaller party, they literally tear their victims to pieces—stomping them, hitting them, biting them, ripping off faces and genitals with their teeth, and sometimes even drinking their blood. If they encounter females with infants, the infants are killed. Sometimes the chimps will attack and even kill females, but by far most killings are reserved for male competitors.[7]

The relationship between organized warfare and mating is a bit more complicated among chimpanzees. Some primatologists have even concluded that patrolling efforts are not for mating purposes per se because males at times attack females and almost never mate with their victims.[8] However, groups successful at systematically killing off males from other troops over time expand their territorial range and increase the number of available females through transfer,[9] that is, through immigration to the larger group. Further, attacks on females are rare, and the characteristics of female targets when they do occur suggest reproductive strategy—younger females in estrus without infants usually endure the least aggression, while females out of cycle with offspring endure the most. This pattern is likely based on the female's reproductive value;[10] younger females without infants make good transfers (and likely good mating prospects over time), whereas older females with infants may compete for food resources and bring potential future rivals.[11]

Not only do females tend to be absorbed by groups successful at exterminating males from neighboring groups,[12] but increased territory from successful raids actually leads to higher reproductive rates among the aggressor group's females.[13] Further, research has found that mating frequencies are positively correlated with the number of patrols.[14]

It is worth mentioning that alpha-male chimpanzees patrol relatively infrequently, whereas middle- to low-ranking males patrol more often.[15] Alphas tend to prefer to mate-guard, as they likely have more to gain, evolutionarily speaking, by staying home with their females than by risking their lives on patrol. It is probably not a coincidence that dominant male humans—such as kings, generals, and presidents—rarely, if ever, go directly into combat alongside the lower-ranking males of their society. In sum, in those species that are closest to us genetically and in which the style of intergroup coalitional aggression most closely resembles our own, sex is inextricable from violent competition for territory.

To observe how these ancient primate scripts play out in humans we have only to look to competitive sports arenas, which spill over with violence, sex, and territory. Take, for example, American football—which in many ways is a metaphor for primate warfare. Here we have groups of males (the players) forming coalitions (teams) to compete for, acquire, and defend territory using aggression. The object of the game is to breach lines of defense and penetrate into enemy territory. As in real warfare, men in this sport break bones and achieve glory, and they generally have enhanced mating prospects—à la team groupies, or access to high-profile females such as actresses or models. They also get paid enormous sums of money to play—ostensibly a reflection of the value we place on a game reenacting ancient scenarios that titillate our evolved design. Lining the playing fields where male coalitions fight for dominance, we find groups of human females in their sexual prime—attractive, scantily clad women *literally* cheering on the violence. Why is it that cheerleaders are rarely ever men, never post-menopausal women, and not ever stifled in Victorian-style outfits? These women brim with sexuality, and their buxom, toned physiques spill over the paltry scraps of fabric purposefully designed to scarcely cover them. This isn't the only sport in which this occurs—combat sports such as boxing or mixed martial arts also follow this tradition, with almost-naked females prancing around the ring before every round of Gombe-esque male-on-male bloody pugilism. Not only is the combination of sexually viable females and male competition a familiar one, but the sports term *score* (or *homerun* for that matter) is often a metaphor in American slang for acquiring sex with a novel woman.

Carry competition forward to the human out-group, and violent competition goes from symbolic to actual killing and from sexual metaphor to mass rape, following the ancient legacy of male mate competition. The idea that rape may have provided evolutionary advantages to male humans has not gone without criticism. But again one must avoid the naturalistic fallacy—that is, the idea that because rape is rooted in the evolutionary past it is good, desirable, or tolerable. Like warfare, it is none of these things, and as a morally detestable act it deserves unflinching examination.

Rape has accompanied warfare in virtually all armed conflicts. Like a firestorm, Genghis Khan (translated as *Universal Lord*) seared his way across Europe, Asia, and Northern Africa, amassing the largest contiguous empire in human history. The philosophy underlying the biggest territorial acquisition known to humankind? "Happiness lies in conquering your enemies, in driving them in front of you, in taking their property, in savoring their despair, in raping their wives and daughters."[16] Following the battle of Okinawa in WWII, it was reported that US troops raped 1,336 women during their first ten days in the city of Kanagawa.[17] When the Japanese army marched across China during WWII, they stopped in Nanking and slaughtered the men before proceeding to rape tens of thousands of Chinese women, including girls, pregnant women, and elderly women. Some estimate the rape toll at between two hundred thousand and eight hundred thousand, leading the massacre to be called the *Rape of Nanking*.[18] Estimates of rape by Russia's Red Army in Berlin alone soar to one hundred thirty thousand, and across Germany to two million.[19] Rape was committed as enthusiastically as killing during the Rwandan genocide:

> One day an official declared, "A woman on her back has no ethnic group." After those words men would capture girls and take them to their fields for sex. Many others feared their wives' reproaches and raped the girls right in the middle of the killing in the marshes, without even hiding from their comrades behind the papyrus.[20]

During the Bosnian War, rape was committed across opposing ethnic forces. The Serbs conducted a massive campaign of rape against Muslim Bosnian girls and women, some as part of their stratagem for ethnic cleansing. Many women were intentionally impregnated and forced to go full-term as a strategy to populate the Bosnian genome with Chetnik blood.[21]

Nor has religious warfare bypassed rape. During the Thirty Years' War (1618–1648), sectarian violence between Christians ravaged Europe, with rivaling denominations hotly competing for territories in Switzerland, Bavaria, Sweden, Germany, Bohemia, Hungary, and France. Rape was pandemic throughout this struggle. Of the scope of this rape Will Durant has remarked that, "armies fed by appropriating the grains and fruits and cattle of the fields . . . and recompensed with the rage to plunder and the ecstasy of killing and rape. . . . The right of a soldier to rape was taken for granted."[22] As David Smith remarks of wartime rape, "Examples could be multiplied indefinitely."[23] Once again we trace the ancient patterns of male primate mate competition in humans—males invading other males' territory, engaging in violent competition, and acquiring sex with resident females.

RAPE AND THE BIBLE

The biblical decrees of God retain a significant legacy in Western notions of morality, and are not only deeply embedded in our moral psychology but also mirrored in many of our legal standards (as in the case of murder and theft and the Ten Commandments). For some, God and religion are in fact synonymous with moral righteousness. For these reasons one might suppose that the Judeo-Christian god and his patriarchal representatives would unequivocally repudiate sexual assault—an illegal act considered morally reprehensible across the world's Christian nations. Not so. Although we have touched upon God-sanctioned rape before, some might be astonished to hear the extent to which these endorsements occur in biblical warfare.

In the story of Judges, men from the tribe of Benjamin tried to kill a Levite man in Gibeah (a hill in or near Jerusalem), and ended up gang-raping and killing his concubine (Judges 19:25). As an act of revenge (here, like men, God punishes the rape of *in-group* members) God instructs the Israelites to march into the territory of Gibeah and slaughter the tribe of Benjamin (Judges 20:21). The Israelites did as they were told and killed twenty-five thousand men. After this they returned and slaughtered the women, children, and animals of Benjamin: "The men of Israel went back to Benjamin and put all the towns to the sword, including the animals and everything else they found. All the towns they came across they set on fire" (Judges 20:48). Six hundred Benjamites survived by retreating to the forests, but all were male soldiers. At some point the Israelites realized that the shortage of women they had created among the Benjamites was threatening the long-term survival of the broader Israelite tribe, which they ultimately sought to preserve. But the Israelites had sworn a sacred oath to God not to give any of their daughters in marriage to the Benjamites, and the oath posed a dilemma. However, the solution was clear to the assembly of patriarchs who formed a war party to invade Jabesh (believed to be east of the river Jordan), murder all the male inhabitants, commit mass infanticide, kill all the women who had ever had sex, and keep all the virgins as spoils of war:

> So the assembly sent twelve thousand fighting men with instructions to go to Jabesh Gilead and put to the sword those living there, including the women and children. "This is what you are to do," they said. "Kill every male and every woman who is not a virgin." They found among the people living in Jabesh-Gilead four hundred young women who had never slept with a man, and they took them to the camp at Shiloh in Canaan. (Judges 21:10–12)

The Israelites gave the stolen virgins to the Benjamites—recall that male chimps perform roughly the same strategy of killing males, infants, and often older females, while letting the younger females live—but there were still not enough women to go around. To address this next impasse, the Israelites instructed the Benjamites to ambush a festival in Shiloh and steal the women festivalgoers:

> "The Benjamite survivors must have heirs," they said, "so that a tribe of Israel will not be wiped out. We can't give them our daughters as wives, since we Israelites have taken this oath: 'Cursed be anyone who gives a wife to a Benjamite.' But look, there is the annual festival of the LORD in Shiloh, which lies north of Bethel, east of the road that goes from Bethel to Shechem, and south of Lebonah." So they instructed the Benjamites, saying, "Go and hide in the vineyards and watch. When

the young women of Shiloh come out to join in the dancing, rush from the vineyards and each of you seize one of them to be your wife. Then return to the land of Benjamin." (Judges 21:17–21)

All of this woman-stealing, of course, really ends up being mass rape. That is, women generally don't have consensual sex with those who murder their entire families, or kidnap them, although there are exceptions to the latter.

The book of Numbers tells another story of revenge, carnage, and rape. God commanded Moses to take vengeance on the people of Midian (believed to be on the shore of Aqaba on the Red Sea), ostensibly for worshipping other gods. The Israelites advanced into Midian territory and did as instructed, killing every man in Midian but capturing women as plunder:

> The Israelites captured the Midianite women and children and took all the Midianite herds, flocks and goods as plunder. They burned all the towns where the Midianites had settled, as well as all their camps. They took all the plunder and spoils, including the people and animals, and brought the captives, spoils and plunder to Moses. (Num. 31:9–12)

Moses, however, was furious. Not for the murder and plunder, but because his men had allowed the women to live. As a compromise, Moses ordered that all the boys and nonvirgins be slaughtered and the virgins be preserved for their uses:

> Now kill all the boys. And kill every woman who has slept with a man, but save for yourselves every girl who has never slept with a man. (Num. 31:17–18).

From here God commands Moses and his priest Eleazar to divide the spoils—including sheep, cattle, donkeys, and *thirty-two thousand virgins*—among the men who fought in battle and other tribal members. They also set apart a percentage of the spoils as tribute to God. That tribute, conveniently for Eleazar, was given to him as God's representative, and to members of Levi's clan who were the caretakers of God's shrine (Num. 31:25–41).

The rape and killing just described proliferated in the biblical age, much as they do in contemporary warfare. When the Israelites marched through the territory of a rival and that group refused to surrender, the Israelites were instructed to appropriate that city. From there God had instructions for the spoils:

> When the LORD your God delivers it into your hand, put to the sword all the men in it. As for the women, the children, the livestock and everything else in the city, you may take these as plunder for yourselves. And you may use the plunder the LORD your God gives you from your enemies. This is how you are to treat all the cities that are at a distance from you and do not belong to the nations nearby. (Deut. 20:13–15).

Further instructions are given unapologetically: "When you go to war against your enemies and the LORD your God delivers them into your hands and you take captives, if you notice among the captives a beautiful woman and are attracted to her, you may take her as your wife" (Deut. 21:10–11). Thus the Bible is straightforward in recommending that men take advantage of territorial conquest to claim nubile women, putting the direction to do so into the mouth of God himself.

Islam was also founded upon territorial gain and sexual ambition. Like the prophets of Judeo-Christianity, Muhammad was himself a dominant male. He had numerous wives, at least one of which, Rayhana, was won in the battle of the Banu Qurayza in which all the men of her tribe, including her husband, were slaughtered. Muhammad was also a skilled military tactician who managed to unite many fractious nomadic tribes of Arabia and their desert gods under one supreme god, consolidating territories across Arabia. The fact that Muhammad was able to accomplish this so rapidly has made him much admired among historians and political scientists—but Muhammad had two (evolutionary) aces in his pocket. Not only were women like Rayhana taken as spoils of war across Muhammad's campaigns, but men who died in battle were also promised a luxuriant male fantasy: a lush garden attended by virgins said to never age (Koran 78:33)—a tempting prize for men programmed to prefer quantity, youthfulness, and parental certainty. Muhammad, it seems, understood the evolved motivations of men.

Over fourteen hundred years later we continue to see groups of young men, ignited by the dominant male imperatives indemnified by their religion, following the same stereotyped paths of primate male mate competition. The "Islamic State" or *ISIS*, a radical group of Islamic militants mostly in Iraq and Syria, have killed their way across the desert territories of the Middle East, beheading their male rivals and capturing thousands of their rivals' women and young girls. After their capture, ISIS raped them, sold them into forced marriages or sexual slavery, and murdered those (many of whom were pregnant) who refused to be sold.[24] They also forced the women belonging to outside religions (such as the Yazidi) to convert to Islam.[25] Based in Nigeria, another militant Islamist group called *Boko Haram* carried out similar strategies in their attempts to convert Nigeria to an Islamic state. Between 2009 and 2014 these men killed over 5000 civilians, mostly males, and kidnapped hundreds of women and schoolgirls.[26] Their name, Boko Haram, translates as *Western education is sin*. By now we have come to understand how education poses a direct threat to despotic male rule. One can presume that part of the perceived threat may also be its close association with the relative sexual independence of Western women, which is so thoroughly at odds with the brand of Islam these young males seek to propagate and with the male sexual dominion it aspires to privilege.

STAKING CLAIM TO MOTHER EARTH

All of the above examples demonstrate that male territoriality can be dangerous. In both secular and religious forms, it brings violence and human suffering. But it holds greater dangers still. While tracts of earth can be conceptualized as the prizes of male mate competition, and the fields of combat as where the vital resources for survival and reproduction are won or lost, the earth is also an intricately balanced ecological system that sustains all life in the biosphere. Collectively, human actions have begun to destroy that balance, and male territoriality plays a key role in this. This is especially true when men, in their relentless drive to compete with other men, seek to control and expend natural resources, ultimately to enact the numbers strategy of reproduction. As we might expect from religious canon already steeped in dominant male privilege, we find that man's style of relating to the natural world is prescribed in the Bible. When such patterns are given divine legitimacy, they carry forward to social policy and become difficult to extricate, further placing humans and all other life-forms in a precarious ecological position. To understand religion's role in ecological ruin, a reframing of territorial space is in order.

The Earth as Ecosystem

An *ecosystem* is a dynamic community of organisms interacting with the inorganic substrate of the earth. Ecosystems usually emerge within specific borders, but many argue that the entire earth is an ecosystem. Ecologists studying ecosystems examine things like soil production, nutrient cycles (how minerals and sun energy are processed through this living web of life), and how the life-giving functions of the earth are balanced and sustained. Plants play an essential role in the web of life. They use photosynthesis to convert radiation from the sun into expendable energy, which is utilized by animals that eat plants (herbivores) and by the carnivorous animals that in turn eat the plant-eaters. Plants also consume carbon dioxide and release oxygen, supporting all terrestrial life forms with breathable air. Just as plants are central to this biosphere, so is the soil that nourishes them. We have come to understand that the fertile earth that men so endeavor to territorialize is actually a vast megalopolis of microscopic, interacting life forms. In his book *The Legacy*, Canadian scientist and environmental activist David Suzuki introduces us to the denizens of soil:

> In a teaspoon of soil we may find hundreds of million to 3 billion bacteria and a million fungi, like yeast and moulds. There is a veritable zoo of creatures in soil, from microscopic fungi, bacteria, yeast, protozoa, rotifers, and roundworms to creatures on the edge of visibility, such as mites and springtails, to the larger woodlice, earthworms, beetles, centipedes, slugs, snails, and ants, and finally to the giants, including moles rabbits, and other rodents. The different groups perform services that keep the soil alive. Bacteria and fungi decompose matter into detritus, which earthworms ingest and excrete as soil nutrient. Worms rummage through the massive amounts of soil, enabling water, air, and organic material to percolate into the matrix.[27]

The more we learn about ecology, the more we understand how systems-level functioning within organismic communities is essential to continued life. When one part of the system is impacted, ripples are felt through the entire system. We are also beginning to understand the detrimental impact of human action on the ecosystem.

Centuries ago, British economist Thomas Malthus (1766–1834)—who influenced Darwin's thinking tremendously—theorized that the population size of a given species will ultimately be kept in check by the environment. When a population increases beyond sustainable levels, it will begin to decline as result of resource depletion and starvation, at which time the surrounding ecosystem's life-giving resources will begin to rebound.[28] Malthus's theory has shown immense reliability over time—with one notable exception. Although humans are bound by biology, their evolved capacity to learn and create technology has been a game-changer in the biological world. Whereas other organisms' adaptations are predicated (principally) by differential survival, humans have managed to create technology that allows them to adapt within the course of a single lifetime. Rather than dying out when food cannot be found, they have managed to master food production like no other animal using various methods of agriculture and animal domestication.

Technology has given humans advantages in other areas, too: rather than developing immunity to disease slowly over generations, humans have created vaccines; rather than slowly developing weapons from their own bodies, they have developed spears, gunpowder, and nuclear arms; rather than developing thick fur, they have mastered fire and invented clothing; and so on. As a result, human populations have veritably exploded around the globe, unchecked and with overwhelming acceleration. To put the growth curve into perspective, consider a statement from the Science Summit on World Populations:

> It took hundreds of thousands of years for our species to reach a population level of 10 million, only 10,000 years ago. This number grew to 100 million people about 2,000 years ago and to 2.5 billion by 1950. Within less than the span of a single lifetime, it has more than doubled to 5.5 billion in 1993.[29]

By March 2012 the human population had swelled to seven billion, with billions more projected into the near future. Scholars from virtually every scientific field have warned that our unbridled, exponential growth is ultimately unsustainable. Humans, though ingenious, are not immune to Malthusian processes. We have managed to commandeer the world's natural resources, but the biosphere remains finite, as do the life-giving assets it contains. In 1798, Malthus made an ominous prediction for humans in his influential work *An Essay on the Principle of Population*:

> The power of population is so superior to the power of the earth to produce subsistence for man, that premature death must in some shape or other visit the human race. The vices of mankind are active and able ministers of depopulation. They are the precursors in the great army of destruction, and often finish the dreadful work themselves. But should they fail in this war of extermination, sickly seasons, epidemics, pestilence, and plague advance in terrific array, and sweep off their thousands and tens of thousands. Should success be still incomplete, gigantic inevitable famine stalks in the rear, and with one mighty blow levels the population with the food of the world.[30]

So far we have managed to skirt Malthus's grave prediction. Technology has allowed us to keep pace with the growing demands of larger and larger human populations, all of which have placed an increasingly heavy burden onto the biosphere. Our agricultural methods have disrupted the earth's nitrogen and phosphorous cycles and polluted our land, rivers, lakes, and oceans, decimating species in the process. Our addictive consumption of fossil fuels has transferred billions of tons of carbon from the earth into the atmosphere, causing global temperatures to rise and polar icecaps to melt. Our insatiable need for space has resulted in deforestation and habitat loss. Overall, it is estimated that human consumption is responsible for the loss of *fifty thousand* species to extinction every year. In place of forests we have built strip malls, freeways, parking lots, and suburban sprawl. By trowelling the world over in concrete, we are slowly beginning to exterminate the very organisms (plants) that provide us with breathable air. All combined, our consumptive primate behaviors are ripping apart the web of life to which we belong and on which we utterly depend for existence.

Male Competition and Resource Consumption

It is important to understand that human population growth is propelled by the growth of ever more complex human economies, the demands of which are met by extracting resources from the natural environment to produce goods and services. For these and other reasons, economic growth (in its current form) has often been pitted against environmental sustainability. Accordingly, human enterprises that regulate economies—such as governments, the financial sector, and corporations—are typically beholden to growth and resistant to any change diverting production and consumption patterns. In theory, the philosophies underwriting economic policy could either hasten or inhibit the destructive environmental impact of economic growth. However, when these policies are influenced by the psychology of male mate competition, we see predictably negative results. To make sense of this influence, we must reflect on how male-typical strategies evolved.

Across the history of our species, homicide and warfare have made male survival an especially tenuous affair for men. Understanding the pressures of male violence helps to illuminate the logic behind patterns of resource acquisition; in an uncertain world, it is a sensible evolutionary strategy to acquire as much territory as you can—and disseminate as many offspring as you can—before you are killed by another male. Territory brings with it survival resources (i.e., food and/or the wealth to ensure a continued food supply), which are used by men as fuel to power their resource-hungry *numbers* strategy of reproduction.

Lest we unfairly place all the blame on men, research shows that women across cultures—seeking survival, status, and stability for themselves and their children—tend to value resource acquisition in potential mates.[31] Accordingly, in environments with higher ratios of men to women, where there is greater mate competition, men save less and incur more debt to make more immediate expenditures and discount large future financial gains for smaller immediate ones.[32] In other words, the greater the level of mate competition, the more willing men are to engage in financial risk to accrue resources likely to increase reproductive success in the short term. Where male sex ratios are higher, women also *expect* men to spend more money on them in their mating efforts.[33] Dominant men, who are more successful at acquiring and expending resources, are more adept at mate competition. There is virtually no ceiling on the benefit of organic resources (or their mediators—financial wealth, status, power, etc.) to male reproductive success, providing enormous incentive for men to continually push the boundaries of economic growth.

When we expand our focus to the scale of nations, there are further connections to be found between male competition and environmental sustainability. Researcher Bryan Husted, for example, has examined how national cultures impact sustainable economic development.[34] In his worldwide study, several cultural dimensions were found to predict the environmental sustainability of economic policy. The details of the research are revealing, particularly the predictor variables, which appear to echo the ideologies of religious dominance.

First, Husted investigated the impact of gendered values on environmental policy, using a masculinity-femininity dimension that differentiates between masculine values based on competitiveness, ambition, power, and materialism and feminine values emphasizing relationships and quality of life. He found that countries endorsing more masculine values showed less social and institutional capacity for environmental sustainability.[35]

Husted also examined the impact of *power distance*, which refers to "the extent to which the less powerful members of institutions and organizations within a country expect and accept that power is distributed unequally."[36] He found that countries with high power-distance were also less likely to espouse environmentally sustainable economic policy.

It should come as no particular surprise that the tenor of religion in a given society tends to vary according to differences in power distance. Religions in low power distance societies tend to stress the equality of believers, whereas religions in high power distance societies are characterized by religious hierarchy.[37] It is also notable that in societies with more feminine values, more women are elected to political office, and religions tend to focus on fellow humans rather than God or gods.[38] This implies that feminine values foster both religious and political ideals based more on equality than on dominance. Recall that religion based on a dominant male god advises men to have "dominion" over the earth and "subdue" it (Gen 1:28), and has all other life forms submitting in fear and dread of men: "The fear and dread of you will fall on all the beasts of the earth, and on all the birds in the sky, on every creature that moves along the ground, and on all the fish in the sea" (Gen 9:2). In this light masculine psychology, and its impact on the religions that undergird the policies of nations, seems to glorify differences in power between men and the rest of the natural world.

Husted goes on to argue that in high-power-distance countries respect for authority suppresses debate about social issues, including those concerning the environment. He adds that high power distance is related to paternalism, where "Paternalism is a system by which superiors provide favors to subordinates in return for their loyalty" and that in such systems, "Decisions are not made on the basis of merit, but on the basis of a balance of favors to subordinates and loyalty to superiors."[39] In other words, systems based more extensively on maintaining alliances with dominants are associated with greater power differentials and reduced environmental sustainability.

Other researchers also found that countries with high power distance and high masculine values, as well as lower educational attainment, show a diminished capacity for environmental sustainability.[40] These researchers argue that the negative relationship of power distance to sustainability is related to the fact that the abuse (or illegal use) of power often goes unchallenged by those with less power.[41] It is not difficult to see how education gives individuals the intellectual basis to challenge power distance and how religions that prohibit questioning—with concepts such as "sinful knowledge"—could be used to undercut any popular challenge to inequitable resource access and use.

Interestingly, the researchers interpreted the impact of high masculine value this way: "Since people in feminine cultures emphasize values, as typical female members do, such as caring for others, interdependence and quality of life, as compared to goal achievement, they tend to care about public goods, including the environment, which is so vital to the well-being of other members in the society."[42] In saying this, they echo a sentiment expressed by the ecofeminist movement, proponents of which would argue that since both women and the natural environment have been "colonized and exploited" by the forces of male dominance, they can more easily establish a sense of unanimity with nature, furthered by the fact that both women and nature create and sustain life. That feminine metaphors for the earth have endured for millennia seems to support this theory, at least in terms of the gendered expectations that such projections imply. Of the power of such comparisons, one feminist philosopher aptly writes:

> In these metaphors, man mediates his engagement with the world through a representation of it as Woman and metaphorically transposes his relation to Woman on to his relation to the world. Many of the metaphors are transcultural and transhistorical. Man speaks of conquering the mountain as he would woman, or raping the land, of his plough penetrating a female earth so that he can sow his seed therein.[43]

If this is true—if men cognitively frame their relationship to the natural environment in a gendered fashion—then patterned masculinity can be targeted in our efforts to reverse the destructive ecological impact of human economies. Because religions based on male competition also impact the environment, they too can serve as points of leverage.

In sum, the research cited above finds that hierarchical, male-driven societies tend to behave in ways detrimental to the worldwide ecosystem. Gods who emphasize power, control, and unquestioning obedience also behave in ways detrimental to the worldwide ecosystem by provoking destructive, domineering approaches to the natural world. This is not to say that religions cause environmental destruction directly. However, once ideologies based on male competition become embedded in religion, they may be difficult to extricate, particularly when hierarchical (male) power is buffered by norms that prohibit questioning.

Religious Rapacity: An Alternate View

Some religious dogmas, such as *man's dominion* and *imago Dei*, are rather explicit in their gendered, domineering approach to the natural world. Man's dominion proclaims man's God-given right to dominate the natural world and expropriate its resources. Once again, God is said to have commanded (italics mine), "Be fruitful and multiply and fill the earth and *subdue it* and have dominion over the fish of the sea and over the birds of the heavens and over every living thing that moves on the earth" (Gen. 1:28). *Imago Dei* gives man the divine right to domineer, as the only creature in the worldwide biosphere created directly in the image of God. Together, *imago Dei* and man's dominion affirm a patterned male approach to the natural world based on consumption and control. In doing so they may inadvertently sacralize an unsustainable economic philosophy and, ultimately, overpopulation.

The Spanish conquest of the Americas exemplifies this ethos of consumptive control. In 1598, conquistador Juan de Oñate—the Spaniard who chopped off the feet of his male Indian rivals at Acoma—formally took possession of the lands of New Mexico. He proclaimed the acquisition in the name of a dominant male God, thus imparting himself with the status of God, declaring his power over life and death and laying claim to every last resource from the seized territory:

> I take and seize tenancy and possession, real and actual, civil and natural, one, two, and three times, one two and three times, one two and three times, and all the times by right I can and should, at this said Rio del Norte, without excepting anything and without limitations, including the mountains, rivers, valleys, meadows, pastures, and waters. In his name I also take possession of all the other lands, pueblos, cities, towns, castles, fortified and unfortified houses which are not established in the kingdoms and provinces of New Mexico, those neighboring and adjacent thereto, and those which may be established in the future, together with their mountains, rivers, fisheries, waters, pastures, valleys, meadows, springs, and ores of gold, silver, copper, mercury, tin, iron, precious stones, salt, morales, alum, and all the lodes of whatever sort, quality or condition they may be, together with the native Indians in each and every one of the provinces, with civil and criminal jurisdiction, power of life and death, over high and low, from the leaves of the trees in the forests to the stones and sands of the river, and from the stones and sands of the river to the leaves in the forests. . . . O holy cross, divine gate of heaven and altar of the only and essential sacrifice of the blood and body of the Son of God, pathway of saints and emblem of their glory, open the gates of heaven to these infidels. Found churches and alters where the body and blood of the Son of God may be offered; open to use a way of peace and safety for their conversion, and give to our king and me, in his royal name, the peaceful possession of these kingdoms and provinces. Amen."[44]

The Spanish not only claimed the riches of the "New World" but extracted them and funneled them back to Spain in a torrent. With this wealth the Spanish crown built plantations, colonies, and cathedrals across the Americas and monopolized the women of those lands, ultimately creating entire continents of *mestizos*, or people of mixed Spanish and Indian blood. Notably, the Spaniards were not alone in their rapacity. The major civilizations they conquered—including the Aztecs, Mayans, and Incas—were also imperialistic societies run by despotic men who were regarded as deities (i.e., societies of high power distance). They too seized land from the surrounding weaker societies and from them extracted gold, jade, corn, chocolate, fruit, game, and women, all in quantities immensely disproportionate to the leftovers they tossed to subjects from the heights of their stone temples. The war gods of the pre-Columbian Americas imparted to their rulers the legitimacy to take from the earth vast riches, just as the Christian god did for the Spanish. Had Mesoamerican civilizations been possessed of the right technologies, they might well be razing the environment in the present day, with the edicts of their male gods fueling the dozers.

When Cortez rode into Mexico City, toppling Aztec shrines and replacing them with cathedrals, he also exchanged Aztec political rulers with those of the Catholic Church and gave the Mexicans a new, dominant-male godhead. Under the aegis of the Christian god, Cortez and the Church discharged their offspring en masse across vast territories—in the genes of a transcontinental Mestizo race and in the minds of converted Catholics. Today, the Catholic Church owns more territory than any multinational corporation in the world. Upon their vast landholdings now tread a vast army of peoples who, not incidentally, call themselves *God's children*. This could not have been accomplished without the driving force of the Church's masculinized psychology, infused as it is with power, ambition, and the lust for material wealth.

Of course, the Church's edict outlawing contraception, with its heavy-handed use of a classic evolutionary *numbers strategy*, has helped, too. Is this really because the Church believes that life starts at fertilization? Like many dominant men before them, leaders of the Church have coded reproductive privilege into law, forbidding contraception for the sake of progeny, power, and economic growth. Mexico City is now the most populated city in the Western hemisphere. The earth beneath it is straining under the weight of its massive population, and the basin atop which the city sits is literally sinking from overuse of the aquifer flowing beneath.

Brazil, another devoutly Catholic nation, harbors grossly overpopulated cities plagued by violent crime and grinding poverty and supports economic policy that is effectively uprooting the Amazonian rainforest (where we get most of our planet's breathable air) in order to feed populations that continue to swell. With the aid of religion, masculine psychology is accelerating humankind precariously closer to a Malthusian end.

Male competition appears—with religion and economics entwined—not only in the *Conquista* but across historical epochs. When these factors appear together, they consistently bring a formula for human overpopulation. The doctrine of Manifest Destiny, for instance, shaped the US government's expansionist policy during the country's conquest of America's western territories. In an 1845 article in the *Democratic Review*, where the term was coined, westward expansion was equated with "the fulfillment of our manifest destiny to overspread the continent allotted by Providence for the free development of our yearly multiplying millions" (with Providence referring, of course, to the guiding hand of God).

Moreover, we continue to see the Abrahamic religions play a role in shaping policy supporting male reproductive prerogatives. In 2012, President Barack Obama mandated that insurance companies cover contraception. The mandate resulted in outcry from American religious factions, resulting in a congressional hearing to debate whether the mandate violated the tenets of religious freedom. Despite the clear impact of contraceptive policy on the lives of women, the congressional panel was comprised of all male religious leaders including a Catholic bishop, a Lutheran reverend, an Evangelical professor of moral philosophy, a rabbi, and a Baptist professor of ethics. Committee chairman Darrell Issa denied the request to have a woman serve on the panel. Representative Carolyn Maloney walked out of the hearing after saying, "I look at this panel, and I don't see one single individual representing the tens of millions of women across the country. . . . Where are the women?"[45]

Once again, it's worth reckoning how strongly our stances on public policy are driven by ideologies tied to our reproductive biology—even though this connection is not immediately apparent. Our ideologies can either drive us forward at full throttle or divert us from a path fraught with suffering. The vision of God as a dominant male that became so much a part of Abrahamic philosophy arose from its utility. His story originated in an age where strong men often raped, plundered, and committed genocide in their ascent to positions of power. In an environment characterized by the utmost brutality, it paid to have a powerful male god who could protect you in battle, reward you with women, and assign all the earth's life-forms to a subordinate status so that you might use them as resources to enact short-term reproductive strategies.

By being uniquely created in God's image, we have become separate from other life-forms, a distancing that, as we have learned from our studies of in-group–out-group psychology, makes it easier for us to destroy them. But in reality the divisions we place between ourselves and the outside world are false, and the ideologies that promote such divisions are unstable over the long term. David Suzuki spoke with the First Nations aboriginal people that he worked with in Canada, whose moral philosophy is now recapitulated by modern science:

> And, they showed me, there is no environment out there and we are here. We are created by the earth by the air we breathe, the water we drink, the food we eat. And, the energy in our bodies, it comes from the sun. We are the environment, whatever we do to the environment, we do directly to ourselves.[46]

Ideology of this kind provides a radical contrast to that of man's dominion and *imago Dei*. Though this First Nations perspective is scientifically sensible in recognizing the very real interdependencies between all life on this planet, I doubt whether this kind of sentiment will take hold on our collective psyche through secular paths alone. Humans are motivated by emotion, and emotions are much more the domain of religion than the cold abstractions of science or philosophy. Faced with impending destruction of the biosphere, and with large-scale violence, there is a need to embrace both secular and religious ideologies that recognize the earth's interdependencies in order to sustain life as we know it. This will require abandoning stances based on the evolutionary imperatives of dominant men, with their long history of conquering territory and exploiting women and their native lands for short-term reproductive gains. Recognizing these patterns is the first crucial step to enacting change.

Chapter 10

RIGHTING OURSELVES

> "Envy thou not the oppressor, and choose none of his ways." —Proverbs 3:31

The great nineteenth-century women's rights movement leader Susan B. Anthony once said, "I distrust those people who know so well what God wants them to do because I notice it always coincides with their own desires."[1] We have seen that historically it has been mostly men claiming to know what God wants and that they inherited their desires from our male primate forebears. God, an omnipotent, immortal being, has been curiously portrayed as having those same desires—demanding access to females and territory—which serve to legitimize the pursuits of powerful men. Meanwhile, men vigorously pursue the social conditions (such as fear, submission, and unquestioning obeisance) that are required for monopolizing such resources. God follows in turn, and his believers are expected to kneel, show fear, obey, and surrender the right to question. Most importantly, men have claimed this dominant male god's backing while perpetrating unspeakable cruelties—including rape, homicide, infanticide, and genocide.

Borrowing from William James, throughout this book I have emphasized the importance of making the "natural seem strange"—of forcing insight into a process that is so natural, so automatized, that it is enacted mostly without conscious awareness. It is precisely because the alpha god paradigm is so intuitive to the primate brain, and so reflexively applied to the hierarchies of religion, that the need for such a book as this exists. While there are potential remedies to the problems engendered by this alpha god, his image is so deeply rooted in our evolved psychology that it is uncertain to what extent solutions can gain traction on a global scale. Boldly unveiling the male primate puppeteers of God will be a necessary first step. Such a move, made introspective and honest by the evolutionary sciences, may help us to develop a more just and compassionate set of religious and secular ethics.

THE PSYCHOLOGY OF THE OTHER

So far we have covered two different kinds of religious violence—that which dominant males use on their subordinates to secure evolutionary rewards and that which in-group members use against religious out-groups. Perhaps the biggest challenge to restraining ourselves from the latter is that it is embedded in our evolved *psychology of the other*, which is among the most deeply rooted and reflexive of all social impulses. For most of human evolution, we evolved in small bands of hunter-gatherers competing for resources with outside bands. Those who were able to cooperate with members of their tribe, in part by forming shared biases against outsiders, had a distinct survival advantage over those who did not. There is even a good case to be made that distorted perceptions of moral fairness (i.e., favoring the in-group) may have been selected for. A growing science of moral cognition is beginning to bring light to these reflexive intergroup biases, making clear that this tribalistic psychology has not gone away. On the contrary, it pervades our moral philosophies, our national politics, and our religions and creates unconscious, irrational biases culminating in hostility against outsiders. Its reflexivity is an indication of just how old that psychology is.

It is so old, in fact, that we share this ancient legacy with contemporary primate species. Jane Goodall observed in her classic *Chimpanzees of Gombe* that "as a result of a unique combination of strong affiliative bonds between adult males on the one hand and an unusually hostile and violently aggressive attitude toward nongroup individuals on the other," the chimpanzee "has clearly reached a stage where he stands at the very threshold of human achievement in destruction, cruelty, and planned intergroup conflict."[2] As for humans, Plato argued that in warfare, Greeks should not enslave other Greeks, burn down their houses, or ravage their fields—these acts should only be performed against *non*-Greeks.[3] Religions often call upon these same instincts to create scriptures that promote in-group–out-group biases. Religion may have even evolved to facilitate in-group cooperation,[4] to include cooperative hostility.

Even so, there are many positive benefits of religion, such as a sense of belonging, existential purpose, and emotional well-being. The Abrahamic scriptures give voice to the best qualities of humanity, such as compassion, justice, honesty, and personal integrity. Below is a brief summary of a far more extensive list:

The Old Testament (Exodus):

- "Honor thy father and thy mother." (20:12)
- "Thou shalt not kill." (20:13)
- "Thou shalt not steal." (20:15)
- "Thou shalt not bear false witness against thy neighbour." (20:16)
- "Though shalt neither vex a stranger, nor oppress him: for ye were strangers in the land of Egypt." (22:21)
- "Ye shall not afflict any widow or fatherless child." (22:22)
- "Do not deny justice to your poor people in their lawsuits." (23:6)

New Testament (1 Peter):

- "Therefore, rid yourselves of all malice and all deceit, hypocrisy, envy, and slander of every kind." (2:1)
- ". . . all of you should be of one mind. Sympathize with each other. Love each other as brothers and sisters. Be tenderhearted, and keep a humble attitude." (3:8)
- "Do not repay evil with evil or insult with insult. On the contrary, repay evil with blessing, because to this you were called so that you may inherit a blessing." (3:9)
- "Offer hospitality to one another without grumbling." (4:9)
- "If you suffer, however, it must not be for murder, stealing, making trouble, or prying into other people's affairs." (4:15)

Koran's Suras:

- ". . . do good. Allah loveth the beneficent." (2:195)
- "Be good to parents and kindred and to orphans and the needy, and speak kindly to mankind." (2:83)
- "Kill not one another." (4:29)

- "There is no good . . . save in him who enjoy almsgiving and kindness and peace-making among the people." (4:114)
- "Be ye staunch in justice, witnesses for Allah, even though it be against yourselves or (your) parents or (your) kindred, whether (the case be of) a rich man or poor man." (4:135)
- "Oh ye who believe! Be steadfast witnesses for Allah in equity, and let not hatred of any people seduce you that ye deal not justly." (5:8)

These suggestions, rules, and laws are simple, compassionate, and morally sensible. Across the globe such edicts have allowed vast groups of people to cooperate and form group identities and a sense of shared moral obligation. But once again, it is an important point to note that such obligations apply almost exclusively to the in-group. Values like kindness, compassion, charity, fairness, and so on tend to drop away at the periphery of many human social groups, and those formed around religion are no exception. Again, on this subject, religious doctrine is clear:

> "O my Lord! For that Thou hast bestowed Thy Grace on me, never shall I be a help to those who sin!" (Koran 28:17)

> "O you who believe! fight those of the unbelievers who are near to you and let them find in you hardness." (Koran 9:123)

> "My son, fear the Lord and the king, and do not join with those who do otherwise." (Prov. 24:21)

> "Whoever will not obey the law of your God and the law of the king, let judgment be strictly executed on him, whether for death or for banishment or for confiscation of his goods or for imprisonment." (Ezra 7:26)

It remains an enormous challenge to extend religious compassion beyond in-group borders. Great thinkers have argued for the value of this vision. Among them, Charles Darwin imagined that:

> As man advances in civilization, and small tribes are united into larger communities, the simplest reason would tell each individual that he ought to extend his social instincts and sympathies to all members of

the same nation, though personally unknown to him. That point being reached, there is only an artificial barrier to prevent his sympathies to all men of all nations and all races.[5]

While some exceptional individuals, upon great introspection, rise above the kinds of insular identifications that make out-group hatred both easy and taken for granted, achieving such insight on a collective level is another matter. We most commonly see these insights expressed by select philosophers, scientists, and meditators, people who spend an extraordinary amount of time in self-reflection and in questioning the basis of human knowledge. In short, it often takes a great deal of focused effort and practice to pull away from the influence of our instincts.

But can metacognition on this scale be realized by the world's mass of religious practitioners? If so, is religious dictum the path forward? Hinduism has voiced this ideal in its teachings, that a wise man "sees himself in all and all in him."[6] But among the major religions of the world such sentiments are rare. And Hindu hands, too, have been stained with the blood of religious violence and oppression. Surely, if the goal is to value all people, the illogical nature of certain religious beliefs may present unique dangers, for example, "God will grant me entrance into heaven if I kill Muslims," or "If I blow myself up (and kill unbelievers in the process) God will give me virgins in paradise."

However, to reiterate, bias against outsiders would exist even without religion. Human beings, like other animals, do not require religion to establish in-group boundaries nor to invade those of other groups. Humans have been known to divide on the basis of race, political ideology, national identity, language, world geography, city-street, and even sports teams. Humans also have a reputation for killing those outside their borders. And religion is not required for one's in-group to be viewed as morally superior. Such biases reflect a long evolutionary history of zero-sum competitions—where one's win is the other's loss—as when two tribes fight for a finite resource. Those competitions are sure to have been a driving force in differential selection, which really means that we are very likely biologically predisposed to have these kinds of biases because on some level doing so provided a survival advantage to our ancestors. Understanding this, then, puts the ultimate blame for religious violence not on religion but on evolution, with religion being just another marker of difference along with politics, street gangs, and sports teams.

Yet religious violence can often seem uniquely glorious. Religion has the capacity to enshroud violence with mystical, sometimes even spiritually ecstatic, experiences whose intensity and intuitiveness ultimately serve to further justify out-group hatred. If you have ever had the misfortune of watching one of the snuff videos released by al-Qaeda in which radical Islamists behead American "infidels" in front of the camera while jubilantly screaming "God is great!," then you know that the sense of glory is revoltingly palpable. What evolutionary theory can do is help us understand why such experiences can feel intuitive and why they are so emotionally powerful. One reason is that when the alpha god paradigm is activated, it engages ancient centers of the brain designed to light up when allying with a more powerful male as we are enthralled with the awe of his power. Further, when a god is portrayed as a dominant male primate, his followers have an instinctively resonant figure to follow into conflict. Alongside this powerful male, believers feel more confident in battle, just as their primate ancestors must have felt for millions of years out on the savage African savannas.

But neither is this to argue that all enmity between human groups *requires* a dominant male. I see the reach and limitations of the dominant male's influence as open questions for which more research is needed. Nevertheless, the impact of dominant males bears reckoning. Any history book can serve as a testament to the mammoth sway that powerful males have held over groups of people. Frequently such men use their influence to enflame their subjects' natural out-group biases, serving to both hasten and intensify out-group violence. Again, these men often stand to benefit greatly from conflict with outside groups, winning outsized shares of territory, sex, and other evolutionary rewards. Men who are better at controlling information, political rhetoric, or religious ideology are more effective in achieving suspicion and hatred of the other tribe, even ensuring that killing the supporters of rival males and their gods becomes seen as proof of in-group commitment, for example: "And slay the Pagans wherever you find them" (Koran 9:5); "We shall cast terror into the hearts of those who disbelieve because they ascribe unto Allah partners, for which no warrant hath been revealed. Their habitation is the Fire" (Koran 3:151).

Dominant men concerned with maintaining their prominence can make questioning the god that has been so instrumental in their rise to power tantamount to treason, an in-group infraction deserving harm or even death. "Lo! those who disbelieve in Allah and His messengers, and seek to make distinction between Allah and His messengers, and say: We believe in some and disbelieve in others, and seek to choose a way in between; Such are disbelievers in truth; and for disbelievers we prepare a shameful doom" (Koran 4:150–51). Power can be won and maintained without the support of a mystical ally, but creating a supremely powerful primate male to appoint the ascendancy of men can help cement the deal—it is an ancient ghost of our past that we humans come psychologically predisposed not to challenge or question.

There are scriptures, allegedly inspired by a dominant male god, that prescribe a sense of personal agency against the momentum of the in-group, for instance: "You must not follow the crowd in doing wrong. When you are called to testify in a dispute, do not be swayed by the crowd to twist justice" (Exod. 23:2). But these passages are few and are engulfed in a roiling sea of edicts urging hatred toward outsiders (or nonbelievers, nonconformers, etc.), and countless other passages in which a dominant male God sets a violent example—as with the god of the Old Testament committing mass infanticide on the Egyptian people (Exodus). In the end, removing the *psychology of the other* from the Bible and the Koran would leave these books in shreds.

The *psychology of the other* is far-reaching in its potential to destroy. With our large brains, not only do we humans possess unsurpassed ability to kill one another, but we have created weapons with the power to fatally eradicate all life in the worldwide biosphere in the process. Collateral damage of this magnitude is historically unrivaled. But this is not the only threat to the natural world. Doctrines based on selfish genes, such as *imago Dei* and man's dominion, deny that we share common origins with the millions of other life-forms on this wondrous blue planet, nurturing a false sense of separation between humans and other beings—in short, forcing all life into a (falsely unrelated) out-group. These doctrines give exclusive privilege to humans in their worldly pursuits, including our preferential right (or duty, as some scriptures) to reproduce. Today the world's fastest growing populations are in the most religious countries.[7] And yet unchecked reproduction, ecological ruin, and violence form a vicious cycle. Population density is related to an increase in male antisocial behaviors, which are often violent.[8] Further, as population swells and resources dwindle, our tribalistic instincts begin to kick in, and we too often revert to making war on our neighbors to survive, much as we have done for all of human history. Edicts that promote a view of the natural world as out-group and that foster unrestrained population growth unwittingly set the conditions for escalating violence.

Despite the destructive power of this man's dominion, the Bible eloquently voices an environmentalist sentiment that is as humble as it is biologically honest:

> That which befalleth the sons of men befalleth beasts; even one thing befalleth them: as the one dieth, so dieth the other; yea, they have all one breath; so that a man hath no preeminence above a beast: for all is vanity. (Eccles. 3:19)

But much more stridence and consistency in propagating this message would be required to divert us from our current path to ecological ruin and the human volatility that comes with overpopulation. How then do we right ourselves and create a less destructive path forward?

The most obvious suggestion might be to outlaw religion, but even if this were indisputably ethical, which it is not, it is unlikely to ever be successful. The need for religion appears to reflect the evolved design of our brains, and the emotional reward of religious experience may be just too powerful. How about adopting nonviolent religions?

PACIFISM AND SELECTIVE OBSERVANCE

Buddhism has been considered an alternative to the more warlike religious divinities. Its more secular forms have even drawn unbelievers, attracted not only by its nonviolent ethical principles but by its acknowledgment of the interconnectivity of all life-forms and perhaps also by the comfort of ritual. Scholars like Robert Wright toy with the idea that the Buddhist practice of mindfulness meditation, with its focus on self-awareness, can serve as a means to transcend the self-centered, emotion-based biases that appear to be at the heart of so much human conflict.⁹ Self-awareness and liberation from the sway of irrational emotions are certainly at the root of Buddhist philosophy. And among all major religions, Buddhist philosophy is perhaps the one most grounded in nonviolence and nonkilling. Another overarching Buddhist theme is to keep one's mind filled with pacifistic and benevolent thoughts, to be mindful of the well-being and interdependence of all beings, large and small, human and nonhuman. Some monks, for example, refrain from walking outside after a rain to avoid stepping on the sentient beings that gather, such as snails or insects.¹⁰ While such a strategy isn't practical for everyone, when properly put into meditative practice, ideologies focused on our intrinsic interconnectivity would seem to counter not only human-on-human conflict, but also that between humans and the entire biosphere.

However, driven by primate ambitions, men will pervert and corrupt any doctrine, and those of Buddhism are no exception. For instance, armed revolts erupted across China for hundreds of years during the middle of the first millennium CE. These wars were driven by ambitious Buddhist monks. Like men from other religious traditions, these monks rose to dominance and martialed large armies to war by making themselves "sons of gods" or the incarnation of the *Maitreya*, a future (male) Buddha.¹¹ Warrior monks littered this period in China and made use of the same patterned methods of manipulation that we have been discussing. One monk by the name of Faqing led an uprising of fifty thousand men using the reliable strategy of jihad, or crusades—when a soldier killed a man he gained the stage of the first *bodhisattva* (roughly, *enlightened being*), when he killed two, the second, and so on, and the more he killed, the closer he got to sainthood.¹²

Like other traditions describing the dominance competitions of gods, certain sutras claim that in one of the Buddha's past lives he slaughtered the Brahmans for heresy.¹³ Thus in the Buddha's concern for heresy we also find the role of male-dominance reputation within Buddhism, just as in the Abrahamic faiths. We also find self-serving moral reasoning. As the story goes, those heretical Brahmans were put to death out of pity for their sins of slandering Buddhism; certain Buddhist doctrines allow for killing if it prevents another from sinning, a kind of compassionate martyrdom, only one that involves killing rather than dying. In other words, "If in killing this man, I go to hell, so be it.

This being must not be doomed to hell."[14] Similarly, certain sutras tell how if a king kills in battle, so long as behind the killing are "compassionate intentions" (e.g., saving a greater number of sentient beings, his children, etc.) he has not accumulated bad karma; the same is said for torture.[15] Thus we can find the blindness of in-group–out-group psychology even in a religion founded on the idea that all sentient beings of the world are part of one universal, inter-related in-group.

As for the dominant ape's compulsion to compete for mates, Buddhist history offers many examples. The Tibetan ruler Songtsan Gampo, for instance, was a warlord king who founded the Tibetan Empire and brought Buddhism to Tibet. Like so many men of the biblical age, he rose to supremacy not only by the sword but by religious mythos rooted in sexual conquest. Legend has it that he subdued Srinmo, the resident earth goddess (or demoness, depending) that spanned Tibetan territory by nailing her to the ground at twelve points on her body. Across Tibet the ambitious king erected stupas—large phallic-shaped, holy structures—which symbolized nails (and other things) penetrating Srinmo. At the center is the Jokhang Temple in Lhasa, the most holy site in Tibetan Buddhism, which is said to represent the "nail" driven into Srinmo's vagina. More to the point, another myth has the male Buddhist god Hayagriva sodomize the male Hindu god Rudra in an act of sexual domination.[16] And so yet again we see territorial acquisition fused with images of sexual conquest and enfolded into the dominant-male-primate machinations of gods—even in Buddhism, arguably the most pacifistic of the great world religions.

That said, it is humans who both create and distort the doctrines of compassion and pacifism, and there is probably much to be learned from the set of moral precepts on which Buddhism is grounded. One could fill many pages with maxims that would seem to counter both the psychology of the other and also the greed of the alpha god paradigm: "Because we all share this small planet earth, we have to learn to live in harmony and peace with each other and with nature. That is not just a dream, but a necessity"[17]; "Life is dear for all. Seeing others as being like yourself, do not kill or cause others to kill"[18]; "If, desiring happiness, you do not use violence to harm living beings who desire happiness, you will find happiness . . ."[19] I do think that certain religious doctrines are more or less pernicious, or more or less humanitarian than others, with one measure being the overall volume of dogmas in one direction or the other. Certainly, most of Buddhist doctrine is explicitly pacifistic, which does not lend itself so easily to co-option by despotic men. Moreover, practices such as mindfulness meditation can be used without adopting any kind of dogma.

Similarly, there is much to be learned from moral precepts drawn from Judaism, Christianity, and Islam. So the question is whether another solution—less drastic than forswearing religion or adopting only those that are explicitly nonviolent—might be to selectively adopt the prosocial edicts within the three great monotheistic religions. But one would have to be pretty selective to avoid also ingesting those passages that serve to sanctify the most heinous of human cruelties. The problem is that the god-kings of the world never want their followers to be selective—they prefer to retain that privilege for themselves. Powerful men throughout history have asserted their own interpretations of religious texts and ensured those interpretations are either accepted in full or rejected at some risk to self. Consider when Jesus reportedly took on what can be interpreted as the whole of Old Testament law:

> For truly I say to you, until heaven and earth pass away, not the smallest letter or stroke shall pass from the Law until all is accomplished. Whoever then annuls one of the least of these commandments, and teaches others to do the same, shall be called least in the kingdom of heaven; but whoever keeps and teaches them, he shall be called great in the kingdom of heaven. For I say to you that unless your righteousness surpasses that of the scribes and Pharisees, you will not enter the kingdom of heaven. (Matt. 5:18–20)

If we consider the Ten Commandments, Sam Harris reminds us that commandments one through four forbid the "practice of any non-Judeo-Christian faith (like Hinduism), most religious art, utterances like 'God damn it!,' and all ordinary work on the Sabbath—all under the penalty of death."[20] Thus to keep the baby, you must actually drink the bathwater—and say it tastes good. Surely, if we were to accept these dogmas as prescribed, most of America would have to be executed.

The Buddha may have recognized the dangers of accepting a dogma totally and unquestioningly. One Buddhist proverb reads: "The holy man is beyond time, he does not depend on any view nor subscribe to any sect; all current theories he understands, but he remains unattached to any of them."[21] Thus not slavishly following a particular ideology, religious or secular, may also be critical for a compassionate spiritual (or secular) ethics. But being selective, at least for the subordinates, is not always easy. It is especially hard under regimes, like those of the Inquisition or Taliban, when armed men stand ready to call any unorthodox lay interpretation an act of heresy.

Thomas Jefferson, one of the architects of the separation of church and state in America, clearly also understood the dangers of dogmatism. In an effort to extract a more rational and ethical set of religious edicts from the Bible, he literally cut and pasted excerpts from Jesus's moral teachings, stripping away any reference to miracles and the supernatural. He describes his intentions in a letter to John Adams dated October 13, 1813, which bears repeating at length:

> In extracting the pure principles which he taught, we should have to strip off the artificial vestments in which they have been muffled by priests, who have travestied them into various forms, as instruments of riches and power to themselves. . . . We must reduce our volume to the simple evangelists, select, even from them, the very words only of Jesus, paring off the amphibologisms into which they have been led, by forgetting often, or not understanding, what had fallen from him, by giving their own misconceptions as his dicta, and expressing unintelligibly for others what they had not understood themselves. There will be found remaining the most sublime and benevolent code of morals which has ever been offered to man. I have performed this operation for my own use, by cutting verse by verse out of the printed book, and arranging the matter which is evidently his, and which is as easily distinguishable as diamonds in a dunghill.[22]

From this passage we see that Jefferson understood how religious teachings could be twisted by men to serve their own dominance pursuits and to more easily serve political ends. It is clear that from all religions we can both extract diamonds and get mired in the dunghills—the latter commonly when dominant-male psychology enters the picture. It is sorting out the difference that can prove most challenging. If we could ever manage to be selective, it would seem a waste to toss aside the positive aspects of religious morality. There are morals and values already in place that if selectively cultivated may help with what Robert Wright has described as "nourishing the seeds of enlightenment indigenous to the world's [moral] tribes."[23] Instead of dispensing with religion, a more feasible solution may be to develop a set of religious ethics that emphasizes the prosocial, the inclusive, and the compassionate and is therefore more reflective of the world in which we currently live, arguably more interdependent, ever-changing, and vulnerable than ever before.

However, another important question is, if a more inclusive religion is feasible, can its ideologies be selectively extracted from the world's extant religions? Admittedly, I am skeptical about this possibility within the Abrahamic faiths, for their books come with built-in proscriptions against selectivity and freethinking, and their god's rule, overall, has the most distinct flavor of totalitarianism. And there is the matter of the beliefs that, while completely illogical, may prove difficult to fling away—talking serpents, virgin birth, resurrection, God-appointed men, rewards in the afterlife for killing, and so on. Certainly Jefferson's version has not gained wide notoriety, despite his stature and visibility.

Some scholars have suggested an overarching morality informed by science and theoretically adoptable by all the world—what philosopher Joshua Greene calls a *metamorality*.[24] But it is not my goal here to attempt to define what this forward-looking spiritual or secular ethos should look like. We are simply not there yet. Kind ideology alone will not prevent ambitious men, backed by millions of years of evolution, from attempting to make subordinates of those around them or from rallying their subordinates to war.

Before any productive discourse can take place, those dominant-male tendencies must be brought to heel, and the right to debate and reimagine must be protected. The most basic of conditions needed for this are freedom of information and separation of church and state. A final support for this kind of dialogue can be gained from teaching the principles of evolution and using the resulting insights to better understand our primate motivations for violence. The right to do this must also be protected, for only in understanding our evolved mind will we be able to create a set of ethics not only adapted for a life on the savanna or among the warrior kings of the biblical age, but for an ever-growing and interdependent community of nations.

ERECTING A WALL

Choosing what to think is a right that dominant men prefer to keep to themselves. Thus one of the most effective means of keeping a populace empowered and protecting it from falling victim to the will of dictators, religious or otherwise, is to ensure the free flow of information in and out of the minds of its members. Freedom of information is essential to keeping governments, and the men who lead them, transparent and accountable. It also enables a populace to expose corruption. Freedom of expression can help to ensure that rules of governance are debated openly and in the best interests of the populace.

Conversely, if we learn anything from history, it should be that censorship is enormously dangerous. Censorship stops truths or ideas from emerging, particularly those that draw attention to inequities of power or wealth, or to abuses by those in positions of authority. Most importantly, censorship keeps a population ignorant, which holds great appeal to dictators. Adolf Hitler, a man responsible for the deaths of an estimated eleven million people, executed a massive book-burning campaign in which science books (by Albert Einstein among them) were widely destroyed across Germany. Moreover, scientists who were Jewish or deemed a threat to Nazi ideology were forbidden to work or publish. They, along with many poets, novelists, and artists, were purged from German society by revoking their citizenship or sending them to their deaths in concentration camps. History is riddled with examples. Joseph Stalin, who was responsible for exterminating some twenty million people, imposed total censorship of all forms of media in the Soviet Union. Mindful of the importance of a dominant-male reputation, he forbade any literature that described defeat of the Soviet Army or fear among its ranks. Given what we have learned, it becomes evident that censorship is really another form of male domination— one unique to our species, whose most important adaptation is the capacity of the mind to process information.

But while leaders need not be religious to rely upon censorship as a tool for domination, neither have religious leaders shied away from this tactic. Popes, inquisitors, imams, and evangelicals have denied freedom of information by burning or forbidding books, and by controlling which books are printed, allowing only one book, or by claiming the exclusive right to interpret books. Freedom of expression has suffered similar injuries. We have seen how words spoken against scripture, gods, or the men who represent gods have too often been answered by censure, rejection, or worse, by torture or killing.

At their most shrewd, religious leaders have also applied tremendous emotional leverage to this kind of intellectual control—by making silence a virtue (calling it faith) and by making questioning a sin (calling it blasphemy). Knowledge itself can be seen as sinful, as is suggested in the central narrative of Adam and Eve and the tree of knowledge. A parallel of knowledge, *reason*, can become the "devil's greatest whore," to borrow from Martin Luther. What religious despots and secular dictators have in common is the understanding that knowledge leads to a sense of empowerment. It is perhaps for this reason that ambitious men have made such an effort to promulgate the central Abrahamic narrative of a terrifying, temperamental god who abhors knowledge and does not suffer questions, particularly those concerning the prevailing power structure. Such men have the advantage of our evolutionary history working in their favor.

This brings us back to one of the most important protections against religious tyranny: the separation of church and state. Thomas Jefferson articulates the point:

> Religious institutions that use government power in support of themselves and force their views on persons of other faiths, or of no faith, undermine all our civil rights. Moreover, state support of an established religion tends to make the clergy unresponsive to their own people, and leads to corruption within religion itself. Erecting the "wall of separation between church and state," therefore, is absolutely essential in a free society.[25]

In 1777, Jefferson drafted the Virginia Statute for Religious Freedom, which was a precursor to the First Amendment. He and other proponents of the separation fully recognized how religious mind control can lead to the abuse of power. Jefferson goes into greater detail:

> Whereas, Almighty God hath created the mind free; That all attempts to influence it by temporal punishments or burthens, or by civil inca-

pacitations tend only to beget habits of hypocrisy and meanness, and therefore are a departure from the plan of the holy author of our religion, who being Lord, both of body and mind yet chose not to propagate it by coercions on either, as was in his Almighty power to do, That the impious presumption of legislators and rulers, civil as well as ecclesiastical, who, being themselves but fallible and uninspired men have assumed dominion over the faith of others, setting up their own opinions and modes of thinking as the only true and infallible, and as such endeavouring to impose them on others, hath established and maintained false religions over the greatest part of the world and through all time; That to compel a man to furnish contributions of money for the propagation of opinions which he disbelieves is sinful and tyrannical.[26]

Now, there is a certain notable irony in the fact that Jefferson argues that God chose not to impose religious coercion despite his power to do so, once again illustrating how powerful men may summon a particular image of God to reinforce their own desires. Similarly, it is fair to say that most of the founding fathers were atheist, or at very least agnostic, and that their views on religious tolerance were to a great extent a practical concern.[27] Jefferson's choice of words was likely intended to engage believers in his efforts to avert religious tyranny. At the heart of these ideas was the prevention of the kinds of religious violence and oppression that had borne such bloody fruit for Europe in the centuries leading up to the revolutionary era.

Rather than outlawing religion outright (as, for example, was attempted in the former Soviet Union), or forcing conformity of religious perspective, Jefferson and his coauthors went on to ground the separation of church and state in America in a more tolerant approach—namely, on the notion that freedom of religious thought and practice and government impartiality in religious matters go hand in hand. Jefferson's corevolutionary, James Madison, addressed this view directly when he wrote that:

Torrents of blood have been spilt in the old world, by vain attempts of the secular arm, to extinguish Religious discord, by proscribing all difference in Religious opinion. Time has at length revealed the true remedy. Every relaxation of narrow and rigorous policy, wherever it has been tried, has been found to assuage the disease.[28]

Two hundred years later, questions remain about this American experiment. Richard Dawkins points out that, "Precisely because America is legally secular, religion has become free enterprise" and illustrates how American churches compete for wealth and congregants in the spiritual marketplace.[29] Some scholars have gone so far as to propose that free-enterprising religions have created unusually high religiosity in America, and while the research on these claims is in dispute,[30] the level of economic power achieved by churches operating in the free market warrants some scrutiny. Churches, and the religious men (and the occasional woman) who run them, have accumulated staggering wealth in America. A surprising number of American evangelists earn million-dollar salaries and own multimillion-dollar mansions and jet airplanes. And to be sure they are a lobbying force to be reckoned with. According to the Pew Research Center, religious advocacy groups, most of which enjoy tax-exempt status, spend hundreds of millions of dollars annually in an effort to influence US public policy.[31] The result is that in America, the wall of separation between church and state that Jefferson described is under constant threat of being breached. Freethinking, freedom of information, and even the freedom to understand the evolutionary designs of our own minds are threatened when the lines between church and state are muddled.

The Texas State Board of Education offers a case study that would be perversely hilarious if it were not such a dangerous example of censorship, which is here entwined with insufficient separation of church and state (as is often the case). For years, the board, which is comprised mostly of members who endorse conservative Christianity, has exuberantly attempted to force creationist teachings into public-school textbooks, censoring evolutionary science and injecting false uncertainty about the veracity of natural selection. Despite the fact that Thomas Jefferson was arguably the most influential political philosopher in American history, credited with writing the Declaration of Independence, the board has attempted to erase him from the history books (perhaps because he is also known for coining the phrase "separation between church and state"). While removing Jefferson from the list of other Enlightenment philosophers—John Locke, Thomas Hobbes, Voltaire, Jean Jacques Rousseau, and Charles de Montesquieu—they added John Calvin,[32] a theologian who, in addition to supporting the murder of heretics, once said, "There is no worse screen to block out the Spirit than confidence in our own intelligence."[33] After much criticism, the board reinstated Jefferson, but removed the term "Enlightenment" from the curriculum.

The board has also attacked the separation of church and state more directly. When one Democratic board member proposed a standard that would have allowed students to "examine the reasons the Founding Fathers protected religious freedom in America by barring government from promoting or disfavoring any particular religion over all others," a Republican member argued that the "founders didn't intend for separation of church and state in America," exclaiming that the statement was "not historically accurate." The board voted down the standard.[34] The fact that these kinds of assaults on freethinking can occur in contemporary America—where both freedom of speech and separation between church and state have legal protection—is a testament to just how insidious religious censorship can be.

In responding to such an assault on truth in education, I am reminded that science remains one of the best means to gather unbiased information. It allows us to test ideas that seem accurate, but that may not be, and to move on from faulty assumptions. Moreover, *it allows us to be skeptical of our own beliefs*. It thus provides an indispensable service to humanity, particularly in light of all the myriad of assumptions that we humans may make based on what feels emotionally intuitive to us. Science allows us to see and acknowledge the evolved design of our brains and to understand how this very design—stunning as it is—leaves our emotions, thoughts, and behaviors so prone to bias.

Despite vilification by an impressively large and economically powerful group of conservative Christians in America, evolutionary science is particularly useful because it can provide unique insight into the ultimate causes of human behavior—especially that which generates human suffering, like religiously motivated violence. At times, evolutionary science may seem off-putting for precisely this reason—because it forces us to acknowledge that we are under certain circumstances predisposed to do terrible things to one another. Some may fear that such an admission diminishes their freedom of choice. But we are not bound to our evolutionary designs; we have the capacity to outsmart them. Our use of contraception is an example often cited by evolutionary scholars, a case in which we have severed the link between our evolved design (to reproduce) and our own desires (to have sex). Wearing eyeglasses or using sunscreen are other examples of our refusal to abide by the genes Mother Nature has dealt us.[35]

The beauty of evolutionary science is that it has the potential to help expand our existing freedoms. When we begin to understand our unconscious evolutionary predispositions, we may begin make more rational and humane choices, such as *not* responding to our destructive impulses, including those imbedded in religious dogma—for instance, scriptures that paint outsiders as inherently immoral, dangerous, or evil. In an environment where evolutionary science is censored by religious powers, the ability for evolutionary knowledge to contribute to a more reflective, more compassionate social dialogue is cut off at the knees. But it can support the nurturance of a more adaptable, humane set of beliefs if allowed a voice in the conversation about who we are. The more we understand about how and why we humans have the tendency to behave the way we do, the greater our potential will be for moving beyond our current limitations.

Strong (and protected) scientific education is not as widely accessible as it should be, nor is training in analytic and critical thinking encouraged in many education systems around the world. And yet it is precisely this kind of critical thinking—this urge to question everything, including why the sky is blue and why God is typically an omnipotent man whose will must be appeased—that will support our continued spiritual and social evolution.

The first principle of education should be to question one's own knowledge and understanding and to strive for impartiality. Here, religious ethics can learn much from scientific ethics. Because science is an impartial tool, scientists (if practicing correctly) strive to follow the dictates of impartiality as closely as possible so as not to contaminate the scientific process with personal bias. It should go without saying that humans can import bias into any endeavor, which scientists acknowledge by their very practice of science; but science has more methods to resist being coopted than most ways of knowing. Conversely, religions are generally very poor at impartiality, owing largely to the fact that dominant men have created gods in their own image and in doing so have made them infallible. By now we should appreciate the danger that such stances pose. To abide by doctrines that cannot be questioned, against which opposing evidence cannot be posed, is to make oneself exquisitely vulnerable to despotic behavior. But science rests on the notion of falsifiability—making assertions that, if false, can be revealed to be false by a particular observation or experiment—in essence, requiring the ability to question. Falsifiability, then, presents great contrast to infallibility. And while still susceptible to the weaknesses of the people who practice it, science starts off with a modicum of humility purposefully built in.

Further, science is adaptable whereas the adaptability of religious doctrine is greatly suspect. When there exists a theory in science about which contrary evidence has been collected through unfettered scientific inquiry, and when enough evidence has been collected to reasonably assume the theory is false, then that theory is abandoned. Thus science is by design open to incoming information. Infallible gods, men, religions, and scriptures, on the other hand, explicitly and intentionally disallow corrective information. They rest on an arrogant sort of self-certainty about the nature of the universe and of the right way to conduct human affairs. Religions have a great deal to say about humility, but as with the rank structures of apes and humans, humility is apportioned unequally according to rank and flows only upwards —humility toward God or dominant religious men is compulsory, but no such modest respect is due to subordinates.

Thus, I suggest that a final principle for fostering spiritual dialogue should be the extension of humility toward knowledge, learning, and other beings. Such humility might follow the basic tenets of science or those rare religious edicts that are based on curiosity about the world and openness to learning new information with the notion that we don't already know all there is to know. Great philosophers across history have taken this approach. Socrates, for instance, who was arguably among the most brilliant of his era, was eager to acknowledge the limits of his understanding—for example, "I know that I have no wisdom, small or great."[36] There have been many other philosophers, scientists, historians, and, even the rare religious leader, who I deem warriors of the greatest spirit, who have taken great personal risk to battle human ignorance and prejudice with rational understanding in an effort to create a more reasoned and compassionate world.

SOCIETAL HEALTH AND FUTURE DIRECTIONS

In keeping with scientific humility, I acknowledge that there is much that we do not yet understand about the science of religious belief and practice. There are many remaining questions. To start with, if religious experience activates reward centers of the brain, then can those experiences be separately stimulated somehow in a manner that doesn't require submission to a more powerful being? Or did our capacity for religion evolve so entwined within the dominance hierarchies of primates that religious ecstasy and "powerful-being worship" are inseparable? If not, could this be done pharmacologically or through genetic engineering or by certain kinds of "spiritual" practice? Meditation may be a way, although it is not clear that can it ever be as emotionally rewarding as the sense of awe that a primate experiences in the presence of a powerful (hypothetical) being. Could such a meditative experience ever be powerful enough to supplant religions based on conquest? Can education push people toward *positive* religious selectivity from doctrines advocating both compassion *and* cruelty? Or do such collections of scriptures in themselves leave an open door for the dominant males of our primate societies to barrel through and enact their evolutionary imperatives, leaving bodies and grief in their wake? I have many questions—and many hypotheses—but more understanding garnered through objective scientific research is crucially needed.

Another question, for which answers are beginning to emerge, is whether secularization results in improving the human condition above and beyond the freedom from oppression or violence at the hands of dominant males. Conservative religious interests often argue that societal health *depends* on belief in God to maintain moral order, social discipline, and just actions. However, it may be useful to look at more secular societies and see how they fare.

Citing research by sociologist Phil Zuckerman, Sam Harris points out that the most secularized nations in the world (such as Denmark, Sweden, Norway, and Holland) fare far better than the most religious countries on a wide array of societal health indices including "life expectancy, infant mortality, crime, literacy, GDP [gross domestic product], child welfare, economic equality, economic competitiveness, gender equality, health care, investments in education, rates of university enrollment, internet access, environmental protection, lack of corruption, political stability, and charity to poorer nations."[37] Other research reaches similar conclusions, showing that more religiosity (including rejecting Darwinian principles) is associated with poorer results on measures of societal health, which leads researchers to conclude that the socioeconomic or existential insecurity that comes with living in less prosperous countries may influence people to glom on to religious belief as a means of coping.[38] In other words, where survival feels less secure, the comfort of religion may have added appeal, but it does not appear to result in creating better conditions of life.

This explanation is compatible both with the research evidence and with a vision of God as the alpha symbol that provides resources and protects against death. But the issue is complex, and a better understanding of the role of dominant-male psychology in religious belief may prove useful here. In the secularized nations on which the research in this area is based, many of the features that make them healthy—such as high gender equality, economic equality, literacy, university enrollment, media access, and environmental protection—would seem to be at odds with dominant-male ambitions. Dominant primate males prefer to monopolize and sexually control females, which in human terms results in poorer gender equality. Among societies run by the most despotic males, socioeconomic equality is only a dream—most attempt to monopolize any indicators of socioeconomic wealth, whether measured in money, the fruits of production, or desired fruits on a Gombean tree. Dominant males have also historically been threatened by knowledge, to which literacy, college enrollment, and media access all lead.

Environmental protections often run into opposition from an economic ideology based on growth-at-all-costs, urged on by the hungry engines of the male reproduction strategy. Thus, while it may be that religiosity per se is counterproductive to societal health, we also cannot discount the influence of the specific religions that predominate in these nations—primarily the Abrahamic faiths with their dominant male god decreeing the rights to oppress women, accumulate wealth, control minds, and destroy the environment to the dominant men that represent him. Lower crime rates may also be traceable to the psychology of dominance. Research indicates that it is not poverty that creates crime, but resource inequality.[39] Unequal resources are almost the invariable result when dominate males begin to monopolize their hierarchical rewards.

Once again the issue is complex and the relationships, understudied, and important questions remain. Is it that dominant men in secularized countries are less able to exert totalitarian rule without the backing of a dominant male god? Does greater secularization empower citizens to shuck off the tenets of religions based on dominance and submission, thus leaving them less vulnerable to ambitious men who would puppeteer God? Does it nourish education or critical thinking skills that help people to meet religion with thoughtful questions? There is a case to be made that people feel less beholden to religious authorities in nations that provide expansive social welfare programs, which once were the domain of the church. Churches were long responsible for welfare, education, health care, orphanages, and so on, but eventually lost their monopolies when universities became the bastions of knowledge and secular state-run social welfare programs began to develop.[40] The least religious and socially healthiest countries also happen to have the best state-run social welfare programs.

In all this research, America emerges as an outlier on many variables. Research has consistently found less religiosity in more economically prosperous countries, with the notable exception of America. However, though exceptionally wealthy, America is also characterized by the greatest economic inequality. It should come as no surprise, then, that in America, religiosity tends to drop away with increasing wealth. Returning to evolutionary psychiatrist John Price and his discussion of the possible adaptations of temporary submission, this raises the question of whether religion in America is used as a strategy to put the subordinate classes in a "giving up state of mind." If so, how much is scripture a factor? There are pervasive ideas in Christianity that would seem to encourage giving up for the possibility of later reward—that Christ will ride down to earth and reverse the fortune of kings in favor of the pious (while razing the entire planet), that the meek shall inherit the earth if they would just wait, and so on. Does this kind of scripture make people more accepting of economic disparity? Does it inhibit their reaction? In mostly atheistic Sweden, the ratio of salaries of CEO's and the general population is 13:1. By contrast, in America, where 80 percent of the citizenry believes in Judgment Day, it's 475:1.[41]

Racism also tends to be associated with religiosity in America.[42] Does having a religion based on a god with a chosen tribe, who espouses genocide against outsiders, play a role in this? Certainly American settlers have a long history of using the Christian religion to dominate the outside races—namely, the African slaves and Native Americans who were made to kneel before Europeans and their dominant male god. What role, if any, do the scriptures of racial domination have on contemporary racism? Consider what advice the Bible offers to slaves:

> Slaves, be obedient to those who are your earthly masters, with fear and trembling, in singleness of heart, as to Christ. (Eph. 6:5)

> Let all who are under the yoke of slavery regard their masters as worthy of all honor so that the name of God and the teaching may not be defamed. (1 Tim. 6:1)

The healthiest countries described above are characterized by great economic equality and all-inclusive social welfare, including paid maternity leave for women, comprehensive medical care, social security, and pensions. America, on the other hand, with its strong cultural emphasis on personal responsibility and entrepreneurship, tends to limit public services and welfare.[43] Freelance researcher Gregory Paul argues that conservative religious political forces in America have supported deregulated and reduced taxation for the rich and notes the irony of how "the religious right that is the main opponent to Darwinian science has become a leading proponent of what has been labeled socioeconomic Darwinism."[44] He goes on to point out that the same forces vigorously promote faith-based charities in place of government-run social services, even though the empirical evidence suggests that faith-based charities are no more effective.[45]

What advantages do the religious conservative forces in America have in keeping the populace reliant on faith-based charities? This, too, is an open question, but there is a cultural history of religious dominance in America in which religious charity was used as a tool for keeping subordinated classes in their social place. For example, African slaves and Native Americans were forced from their former means of self-sufficiency—in the former by taking them from their native lands and in the latter by eradicating the animals on which they subsisted and appropriating the land on which they dwelled. In return they were yoked by the forced charity of the churches and missions that administered both their food and their indoctrination into the Christian religion, much through selectively proselytizing the doctrines of submission noted above. This strategy is not new to America. Napoleon once said that:

> Society cannot exist without the inequality of fortunes, and inequality of fortune cannot exist without religion. When one man is dying of hunger by the side of another who is overfed, it is impossible for him to submit to this difference unless there is an authority which says to him: "God wills it thus: there must be rich and poor in the world, but afterwards and for all eternity matters, will be otherwise arranged."[46]

How much, if any, do the doctrines of submission in Christianity relate to the vast inequities of wealth in America? And why is it that the wealthy in America are less likely to be religious? Does having power make you less vulnerable to doctrines of submission? If so by what means?

Finally, in constricting the room for powerful males to corrupt the societal health of nations, there may be a corresponding value to empowering women. While women tend to be attracted to dominant traits in males and reward dominant men with sex (likely the most powerful of all motivators for males to dominate), the extent to which women's own needs are subjugated depends on their economic independence from men. For instance, the societies that least emphasize chastity (a form of female sexual control) —again, seen most in Scandinavia—are the ones that provide the best social services for women, such as paid childcare, extended paid maternity leave, and other material benefits. As David Buss explains, "Where women control their economic fate, do not require so much of men's investment, and hence need to compete less, women are freer to disregard men's preferences. . . . Men everywhere might value chastity if they could get it, but in some cultures they simply cannot demand it from their brides."[47]

The extent to which religious dogma influences public policy around female sexuality and economic power is another open question. But consider the potential power of messages in the Bible, such as in Deuteronomy, in which a man who discovers his bride is not a virgin is allowed to have her stoned to death at her father's doorstep (Deut. 22:13–21). Here there is (or should be) obvious ethical value to returning to women the control of their sexuality and support for such a move through public policy.

However, the take-home message from this line of research is that less religion is associated with better societal health, which is precisely contrary to the argument made by conservative Christians in America and suggests once more the potential ills of relying too heavily on an ethics based on appeasing the whims of an alpha god.

CLOSING THOUGHTS

I am not among those who think that science can ever replace religion, at least not on a global scale. Moreover, I have a deep appreciation for the sense of meaning that acknowledging being a part of something greater than ourselves can provide. But I reject the idea that that something is a supernatural dominant male and that we are made in his image, connected to him by lineal descent or as subordinates in the hierarchy. Such characterizations reflect our limitations as animals with brains designed to navigate social hierarchies, far too much to bear up under close scrutiny. I also understand how frightening it can be to admit that we are not special, protected beings with hope of eternal life and to view ourselves as animals, related to apes, ultimately doomed to the same impermanence as all other beings. Nevertheless, I, like many other scientists, find a sense of awe and wonder in the fact that we are far more interconnected with the other life-forms on this planet than we once thought, which is indeed something greater than our individual selves. And truly, this realization indicates that we are not quite what we seem.

We are not beings anchored in time as we imagine ourselves to be. We are part of a continuous flow through rivers of space and time—vast streams of ancestors and of the genes flowing within them, each of us but one moment in the current. Science reveals to us a picture in which our bodies and brains are shaped by our genetic programming, our sense of self is a product of our brain's activity, and our brains are built incrementally upon not only our own experiences but also the culmination of all the brains that have thought before us. As a result we see through the eyes of our ancestors. Each of us does things, loves things, abhors things, thinks things, and aspires to things as amalgams of all the life that has preceded our own existence, stretching our very experience of being across the generations. To my mind, recognizing that we come from apes is a part of this, which is something extraordinary. My sense of awe comes not merely from our relationship to apes but to all other life-forms on this planet, and also from the understanding that each human is a virtual universe of networking life-forms unto herself. There are innumerable ways in which the natural world can elicit awe and mystery, rendering questionable the need for a supernatural one; this wonder only increases when we understand and accept our place within it.

ACKNOWLEDGMENTS

Jen Kuzara: Your feedback was especially precise and your feminist perspectives, extremely valuable. Benjamin Purzycki: Thank you for your thoughtful input from cover to cover, for your encouragement, and for our engaging conversations.

Erin: No one helped me as much or in so many ways as you. Your editing was not only tireless and brilliant, but given with such inspiration that you made the strenuous process of writing a book a breathtaking, contemplative journey. This book, and the growth it afforded me, would not be possible without you.

My father: You sat me down at a very young age and taught me biology, evolution, and defiant intellectualism. Your history, your heritage, and your resistance to the abuse of power live on in the pages of this book.

Lastly, I thank all the great thinkers who have come before. You form the bedrock on which all of science is built. Your fire, your vision, and your great personal risks are not forgotten.

NOTES

CHAPTER 1: ENTER GOD THE DOMINANT APE

1. The Pew Research Center, "The Global Religious Landscape," December 18, 2012. http://www.pewforum.org/2012/12/18/global-religious-landscape-exec/ (accessed November 10, 2014).

2. B. M. Knauft, "Violence and Sociality in Human Evolution," *Current Anthropology* 32, no. 4 (1991).

3. F. de Waal, "Integration of Dominance and Social Bonding in Primates," *Quarterly Review of Biology* 61 (1986) (as cited in Knauft, "Violence and Sociality in Human Evolution").

4. F. J. White, "Social Organization of Pygmy Chimpanzees," *Understanding Chimpanzees*, ed. P. G. Heltne and L. A. Marquardt (Cambridge, MA: Harvard University Press, 1989) (as cited in Knauft, "Violence and Sociality in Human Evolution").

5. M. N. Muller and J. C. Mitani, "Conflict and Cooperation in Wild Chimpanzees," *Advances in the Study of Behavior* 35 (2005).

6. N. A. Chagnon, *Yąnomamö: The Fierce People*, 3rd ed. (New York: Holt, Rinehart & Winston, 1983).

7. J. Henrich and F. J. Gil-White, "The Evolution of Prestige: Freely Conferred Deference as a Mechanism for Enhancing the Benefits of Cultural Transmission," *Evolution and Human Behavior* 22 (2001): 167.

8. Ibid., p. 165.

9. R. N. Bellah, *Religion in Human Evolution: From the Paleolithic to the Axial Age* (Cambridge, MA: Belknap Press of Harvard University Press, 2011); and W. Johnson and T. Earl, *The Evolution of Human Societies*, 2nd ed. (Stanford, CA: Stanford University Press, 2000).

10. R. Wright, *The Evolution of God* (New York: Little, Brown, 2009).

11. Ibid., p. 19.

12. S. Gaschet, *The Klamath Indians of Southwestern Oregon* (Washington, DC: US Government Printing Office, 1890) (as cited in ibid.).

13. L. Marshall, "!Kung Bushman Religious Beliefs," *Africa: Journal of the International African Institute* 32 (1962) (as cited in Wright, *Evolution of God*).

14. Johnson and Earl, *Evolution of Human Societies*, and Wright, *Evolution of God*.
15. Wright, *Evolution of God*, p. 59.
16. S. L. Rogers, *The Shaman: His Symbols and His Healing Power* (Springfield, IL: Charles C. Thomas, 1982) (as cited in Wright, *Evolution of God*).
17. A. Watson, *The Evolution of International Society* (New York: Routledge, 1992), p. 26 (as cited in Wright, *Evolution of God*).
18. Wright, *Evolution of God*, p. 82.
19. Hammurabi, *Code of Hammurabi* (Rockville, MD: Wildside, 2009), p. 7.
20. J. Bottero, *Religion in Ancient Mesopotamia* (Chicago: University of Chicago Press, 2000) (as cited in Wright, *Evolution of God*).
21. Wright, *Evolution of God*, p. 88.
22. W. Durant, *Our Oriental Heritage* (New York: Simon and Schuster, 1935).
23. Ibid., p. 233.
24. Wright, *Evolution of God*, p. 139.
25. Johnson and Earl, *Evolution of Human Societies*.
26. Wright, *Evolution of God*, p. 139.
27. Ibid.
28. Ibid.

CHAPTER 2: EVOLUTIONARY MECHANISMS: ETIOLOGY

1. J. Archer, "Does Sexual Selection Explain Human Sex Differences in Aggression?" *Behavioral and Brain Sciences* 32, nos. 3–4 (2009): 249–66.
2. N. A. Chagnon, *Yąnomamö: The Fierce People*, 3rd ed. (New York: Holt, Rinehart & Winston, 1983).
3. D. M. Buss, "The Multiple Adaptive Problems Solved by Human Aggression," *Behavior and Brain Science* 32, nos. 3–4 (2009): 271. This article is a commentary on Archer, "Does Sexual Selection Explain Human Sex Differences in Aggression?"
4. The deer and peacock analogies that I use came from my reading of D. M. Buss, "Multiple Adaptive Problems."
5. R. A. Fisher, *The Genetical Theory of Natural Selection* (Oxford: Oxford University Press, 1930).
6. F. de Waal, *Chimpanzee Politics: Power and Sex among Apes* (Baltimore, MD: Johns Hopkins University Press, 1998), p. 25.
7. B. B. Smuts, *Sex and Friendship in Baboons* (New York: Aldine, 1985).
8. Ibid.; D. M. Buss, *The Evolution of Desire: Strategies of Human Mating* (New York: Basic Books, 1994).

9. J. S. Gillis and W. E. Avis, "The Male-Taller Norm in Mate Selection," *Personality and Social Psychology Bulletin* 6, no. 3 (1980).

10. B. Pawlowski, R. I. M. Dunbar, and A. Lipowicz, "Tall Men Have More Reproductive Success," *Nature* 403 (2000).

11. B. Pawlowski and G. Jasienska, "Women's Preferences for Sexual Dimorphism in Height Depend on Menstrual Cycle Phase and Expected Duration of Relationship," *Biological Psychology* 70, no.1 (2005).

12. S. Pinker, *How the Mind Works* (New York: W. W. Norton, 1997).

13. F. A. Beach and L. Jordan, "Sexual Exhaustion and Recovery in the Male Rat," *Quarterly Journal of Experimental Psychology* 8, no. 3 (1956).

14. R. P. Michael and D. Zumpe, "Potency in Male Rhesus Monkeys: Effects of Continuously Receptive Females," *Science* 200, no. 4340 (1978).

15. Buss, *Evolution of Desire*.

16. Ibid., p. 78.

17. Ibid., p. 79.

18. D. M. Buss et al., "Sex Differences in Jealousy: Evolution, Physiology, and Psychology," *Psychological Science* 3, no. 4 (1992); B. P. Buunk et al., "Sex Differences in Jealousy in Evolutionary and Cultural Perspective: Tests from the Netherlands, Germany, and the United States," *Psychological Science* 7, no. 6 (1996); D. A. DeSteno and P. Salovey, "Evolutionary Origins of Sex Differences in Jealousy: Questioning the 'Fitness' of the Model," *Psychological Science* 7, no. 6 (1996).

19. Buss, *Evolution of Desire*.

20. D. Singh and P. M. Bronstad, "Female Body Odour Is a Potential Cue to Ovulation," *Proceedings of the Royal Society of London B* (2001) (as cited in Buss, *Evolution of Desire*).

21. W. S. Gangestad, R. Thornhill, and E. C. Garver, "Changes in Women's Sexual Interests and Their Partners' Mate-Retention Tactics across the Menstrual Cycle: Evidence for Shifting Conflicts of Interest," *Proceedings of the Royal Society of London B* 269, no. 1494 (2002) (as cited in Buss, *Evolution of Desire*).

22. Buss, *Evolution of Desire*, p. 25.

23. W. D. Hamilton, "The Genetical Evolution of Social Behaviour I, II," *Journal of Theoretical Biology* 7, no. 1 (1964).

24. M. E. Hauber and P. W. Sherman, "Self-Referent Phenotype Matching: Theoretical Considerations and Empirical Evidence," *Trends in Neurosciences* 24, no. 10 (2001).

25. R. Dawkins, *The Selfish Gene* (Oxford: Oxford University Press, 1976).

26. J. Ackerman, *Chance in the House of Fate: A Natural History of Heredity* (New York: Mariner, 2001), p. 141.

27. L. Cosmides and J. Tooby, "Evolutionary Psychology: A Primer," http://www.psych.ucsb.edu/research/cep/primer.html (accessed May 5, 2012).

28. Ibid.

29. O. Devinsky and G. Lai, "Spirituality and Religion in Epilepsy," *Epilepsy & Behavior* 12, no. 4 (2008).

30. S. Baron-Cohen, A. M. Leslie, and U. Frith, "Does the Autistic Child Have a 'Theory of Mind?'" *Cognition* 21, no. 1 (1985).

31. Ibid.

32. D. Johnson and J. Bering, "Hand of God, Mind of Man: Punishment and Cognition in the Evolution of Cooperation," *Evolutionary Psychology* 4 (2006).

33. P. Boyer, "Are Ghost Concepts 'Intuitive,' 'Endemic,' and 'Innate?'" *Journal of Cognition and Culture* 3 (2003).

34. B. G. Purzycki and R. Sosis, "The Extended Religious Phenotype and the Adaptive Coupling of Ritual and Belief," *Israel Journal of Ecology & Evolution* 59, no. 2 (2013).

35. S. Guthrie, *Faces in the Clouds* (Oxford: Oxford University Press, 1993).

36. J. Teehan, *In the Name of God: The Evolutionary Origins of Religious Ethics and Violence* (Chichester, UK: Wiley-Blackwell, 2010).

37. J. L. Barrett, *Why Would Anyone Believe in God?* (Lanham, MD: AltaMira, 2004).

38. B. J. Scholl and P. Tremoulet, "Perceptual Causality and Animacy," *Trends in Cognitive Sciences* 4, no. 8 (2000).

39. P. Boyer, *Religion Explained: The Evolutionary Origins of Religious Thought* (New York: Basic Books, 2001).

40. C. F. Zink et al., "Know Your Place: Neural Processing of Social Hierarchy in Humans," *Neuron* 58, no. 2 (2008).

41. A. Moors and J. de Houwer, "Automatic Processing of Dominance and Submissiveness," *Experimental Psychology* 52, no. 4 (2005).

CHAPTER 3: THE PROTECTOR GOD

1. D. L. Smith, *The Most Dangerous Animal: Human Nature and the Origins of War* (New York: St Martin's Griffin, 2007).

2. P. Shepherd, *The Others: How Animals Made us Human* (Washington, DC: Island/Shearwater, 1996), p. 29 (as cited in ibid.).

3. J. A. Byers, *American Pronghorn: Social Adaptations & the Ghosts of Predators Past* (Chicago: University of Chicago Press, 1997) (as cited in Smith, *Most Dangerous Animal*).

4. C. R. Darwin, "A Biographical Sketch of an Infant," *Mind: A Quarterly Review of Psychology and Philosophy* 2, no. 7 (1877): 288.

5. S. Agras, D. Sylvester, and D. Oliveau, "The Epidemiology of Common Fears and Phobias," *Comprehensive Psychiatry* 10, no. 2 (1969).

6. S. B. Hrdy, *The Woman that Never Evolved* (Cambridge, MA: Harvard University Press, 1999).

7. E. Marais, *My Friends the Baboons* (London: Blond and Briggs, 1975).

8. M. Hiraiwa-Hasegawa et al., "Aggression toward Large Carnivore by Wild Chimpanzees of Mahale Mountains National Park, Tanzania," *Folia Primatologica* 47, no. 1 (1986).

9. R. B. Lee, *The !Kung San: Men, Women and Work in a Foraging Society* (New York: Cambridge University Press, 1979).

10. T. N. Headland and H. W. Green, "Hunter-Gatherers and Other Primates as Prey, Predators, and Competitors of Snakes," *Proceedings of the National Academy of Sciences USA* 108, no. 52 (2011).

11. G. Pinch, *Egyptian Mythology: A Guide to the Gods, Goddesses, and Traditions of Ancient Egypt* (Oxford: Oxford University Press, 2002).

12. Euripides, *Herakles*, trans. Tom Sleigh (New York: Oxford University Press, 2001).

13. E. Becker, *The Denial of Death* (New York: Free Press, 1973), p. 26.

14. M. J. Landau et al., "Deliver Us from Evil: The Effects of Mortality Salience and Reminders of 9/11 on Support for President George W. Bush," *Personality and Social Psychology Bulletin* 30, no. 9 (2004).

15. H. Fineman, "The Gospel According to George," *Newsweek*, April 25, 2004, http://www.newsweek.com/gospel-according-george-125363 (accessed September 20, 2011).

16. E. Becker, *The Birth and Death of Meaning*, 2nd ed. (New York: Free Press, 1971), p. 161.

17. "BBC Profile: Nicaraguan President Daniel Ortega," *BBC*, November 6, 2011, http://www.bbc.co.uk/news/world-latin-america-15544315 (accessed February 20, 2012); H. Williams, "Violeta Barrios de Chamorro," in *Women in World Politics: An Introduction*, by F. D'Amico, ed. P. R. Beckman (London: Bergin and Garvey, 1995).

18. A. C. Little et al., "Facial Appearance Affects Voting Decisions," *Evolution and Human Behavior* 28 (2007); D. E. Re et al., "Facial Cues to Perceived Height Influence Leadership Choices in Simulated War and Peace Contexts," *Evolutionary Psychology* 11, no. 1 (2013).

19. B. R. Spisak et al., "Warriors and Peacekeepers: Testing a Biosocial Implicit Leadership Hypothesis of Intergroup Relations Using Masculine and Feminine Faces," *PLoS ONE* 7, no. 1 (2012).

20. A. Norenzayan and I. G. Hansen, "Belief in Supernatural Agents in the Face of Death," *Personality and Social Psychology Bulletin* 32, no. 2 (2006).

21. M. Osarchuk and S. J. Tatz, "Effect of Induced Fear of Death on Belief in Afterlife," *Journal of Personality and Social Psychology* 27, no. 58 (1973): 308–318; R. Willer, "No Atheists in Foxholes: Motivated Reasoning and Religious Ideology," in *Social and Psychological Bases of Ideology and System Justification*, ed. J. T. Jost, A. C. Kay, and H. Thorisdottir (Oxford: Oxford University Press, 2009).

22. J. C. Buchan et al., "True Paternal Care in a Multi-Male Primate Society," *Nature* 425, no. 6954.

23. C. Borries et al., "Males as Infant Protectors in Hanuman Langurs (Presbytis entellus) Living in Multi-Male Groups: Defence Pattern, Paternity and Sexual Behaviour," *Behavioral Ecology Sociobiology* 46 (1999).

24. Pope Benedict XVI, "Audience: What it means to call God 'Father,'" *Official Vatican Network: Vatican Radio*, http://www.news.va/en/news/audience-what-it-means-to-call-god-father (accessed July 12, 2013).

25. L. Parr and F. de Waal, "Visual Kin Recognition in Chimpanzees," *Nature* 399 (1999).

26. S. Platek et al., "How Much Paternal Resemblance is Enough? Sex Differences in Hypothetical Investment Decisions, but not in the Detection of Resemblance," *Evolution and Human Behavior* 24, no. 2 (2003).

27. S. Platek et al., "Reactions towards Children's Faces: Resemblance Matters More for Males than Females," *Evolution and Human Behavior* 23, no. 3 (2002).

28. A. Alvergne, C. Faurie, and M. Raymond, "Father–Offspring Resemblance Predicts Paternal Investment in Humans," *Animal Behaviour* 78, no. 1 (2009).

29. M. J. E. Charpentier et al., "Message 'Scent': Lemurs Detect the Genetic Relatedness and Quality of Conspecifics via Olfactory Cues," *Animal Behaviour* 80, no. 1 (2010).

30. T. Aquinas, *Summa Theologica*, vol. 1, pt. 1, trans. Fathers of the English Dominican Province (New York: Cosimo Classics, 2007).

31. Ibid., p. 470.

32. F. B. M. de Waal, "The Organisation of Agonistic Relations within Two Captive Troops of Java Monkeys (Macaca fascicularis)," *Zeitschrift für Tierpsychologie* 44 (1977); B. Chapais, "The Role of Alliances in Social Inheritance of Rank among Female Primates," in *Coalitions and Alliances in Humans and Other Animals*, vols. 29–59, ed. S. A. Harcourt and F. B. M. de Waal (Oxford: Oxford Science Publications, 1992); J. B. Silk, A. Samuels, and P. S. Rodman, "The Influence of Kinship, Rank, and Sex upon Affiliation and Aggression among

Adult Females and Immature Bonnet Macaques (Macaca radiata)," *Behaviour* 78, no. 1/2 (1981).

33. C. P. van Schaik and J. Janson, eds., *Infanticide by Males and Its Implications* (Cambridge: Cambridge University Press, 2000).

34. C. P. van Schaik and P. M. Kappeler, "Infanticide Risk and the Evolution of Male-Female Association in Primates," *Proceedings of the Royal Society B* 264 (1997).

35. M. Daly and M. Wilson, "Some Differential Attributes of Lethal Assaults on Small Children by Stepfathers versus Genetic Fathers," *Ethology and Sociobiology* 15, no. 4 (1984).

36. J. Pelikan, ed., *Luther's Works* (St. Louis: Concordia Publishing House, 1958).

37. D. P. Watts, "Reciprocity and Interchange in the Social Relationships of Wild Male Chimpanzees," *Behaviour* 139, no. 2/3 (2002).

38. D. P. Watts, "Grooming Between Male Chimpanzees at Ngogo, Kibale National Park. II. Influence of Male Rank and Possible Competition for Partners," *International Journal of Primatology* 21, no. 2 (2000); K. Arnold and A. Whiten, "Grooming Interactions among the Chimpanzees of the Budongo Forest, Uganda: Tests of Five Explanatory Models," *Behaviour* 140, no. 4 (2003).

39. D. P. Watts, "Coalitionary Mate Guarding by Male Chimpanzees at Ngogo, Kibale National Park, Uganda," *Behavioral Ecology and Sociobiology* 44 (1998).

40. U. Gerloff et al., "Intracommunity Relationships, Dispersal Pattern and Paternity Success in a Wild Living Community of Bonobos (Pan paniscus) determined from DNA Analysis of Faecal Samples," *Proceedings of the Royal Society B* 266, no. 1424 (1999); A. Widdig et al., "A Longitudinal Analysis of Reproductive Skew in Male Rhesus Macaques," *Proceedings of the Royal Society B* 271, no. 1541 (2004); B. J. Bradley et al., "Mountain Gorilla Tug-of-War: Silverbacks have Limited Control over Reproduction in Multi–Male Groups," *Proceedings of the National Academy of Sciences* 102, no. 6 (2005).

41. L. L. Betzig, *Despotism and Differential Reproduction: A Darwinian View of History* (New York: Aldine, 1986).

42. W. Burkert, *Creation of the Sacred: Tracks of Biology in Early Religions* (Cambridge, MA: Harvard University Press, 1996), p. 95.

43. W. Durant and A. Durant, *The Age of Faith: A History of Medieval Civilization (Christian, Islamic, and Judaic) from Constantine to Dante, AD 325–1300*, vol. 4 (New York: Simon and Schuster, 1950).

44. P. de Rosa, *Vicars of Christ: The Dark Side of the Papacy* (Dublin: Poolbeg, 2000).

45. Durant and Durant, *Age of Faith*.

46. B. R. Lewis, *A Dark History: The Popes: Vice, Murder, and Corruption in the Vatican* (New York: Metro, 2009).

47. Ibid.

48. W. Durant and A. Durant, *The Age of Napoleon*, vol. 11 (New York: Simon and Schuster, 1975).

49. C. Esdaile, *Napoleon's Wars: An International History* (New York: Penguin Group, 2007), p. 185.

50. A. Kamen, "George W. Bush and the G-Word," *Washington Post*, October 14, 2005, http://www.washingtonpost.com/wp-dyn/content/article/2005/10/13/AR2005101301688.html (accessed November 23, 2011).

51. J. Borger, "How Born-Again George Became a Man on a Mission," *The Guardian*, October 6, 2005, http://www.theguardian.com/world/2005/oct/07/usa .georgebush (accessed November 23, 2011).

CHAPTER 4: SEXUAL DOMINANCE: FROM APES TO MEN TO GODS

1. D. C. Geary, J. Vigil, and J. Byrd-Craven, "Evolution of Human Mate Choice," *The Journal of Sex Research* 41, no. 1 (2004).

2. D. Fossey, *Gorillas in the Mist* (New York: First Mariner, 1983).

3. F. de Waal, *Chimpanzee Politics: Power and Sex among Apes* (Baltimore, MD: Johns Hopkins University Press, 1998), p. 163.

4. D. Maestripieri, *Macachiavellian Intelligence: How Rhesus Macaques and Humans Have Conquered the World* (Chicago: University of Chicago Press, 2007).

5. Ibid.

6. D. Maestripieri et al., "One-Male Harems and Female Social Dynamics in Guinea Baboons," *Folia Primatol* 78 (2007).

7. J. J. Abegglen, *On Socialization in Hamadryas Baboons* (London: Associated University Presses, 1984); R. I. M. Dunbar, "The Social Ecology of Gelada Baboons," in *Ecological Aspects of Social Evolution: Birds and Mammals*, ed. D. I. Rubenstein and R. Wrangham (Princeton: Princeton University Press, 1986).

8. D. Maestripieri et al., "One-Male Harems and Female Social Dynamics in Guinea Baboons;" L. Swedell and A. Schreier, "Male Aggression towards Females in Hamadryas Baboons: Conditioning, Coercion, and Control," in *Sexual Coercion in Primates: An Evolutionary Perspective on Male Aggression against Females*, ed. M. Mueller and R. Wrangham (Cambridge, MA: Harvard University Press, 2009).

9. A. F. Dixson, *Primate Sexuality: Comparative Studies of the Prosimians, Monkeys, Apes, and Humans*, 2nd ed. (Oxford: Oxford University Press, 2012).

10. T. Furuichi, "Agonistic Interactions and Matrifocal Dominance Rank of Wild Bonobos (Pan paniscus) at Wamba," *International Journal of Primatology* 18 (1997).

11. T. Kano, "Male Rank Order and Copulation Rate in a Unit-Group of Bonobos at Wamba, Zaïre," in *Great Ape Societies*, ed. W. C. McGrew, L. F. Marchant, and T. Nishida (Cambridge: Cambridge University Press, 1996).

12. F. de Waal, "The Brutal Elimination of a Rival among Captive Male Chimpanzees," in *Ostracism: A Social and Biological Phenomenon*, ed. M. Gruter and R. D. Masters (Amsterdam: Elsevier, 1986).

13. C. Borries et al., "DNA Analyses Support the Hypothesis that Infanticide is Adaptive in Langur Monkeys," *Proceedings of the Royal Society B* 266 (1999); C. van Schaik and J. Janson, eds., *Infanticide by Males and Its Implications* (Cambridge: Cambridge University Press, 2000); C. Borries et al., "Males as Infant Protectors in Hanuman Langurs (Presbytis entellus) Living in Multi-Male Groups: Defence Pattern, Paternity, and Sexual Behaviour," *Behavioral Ecology and Sociobiology* 46 (1999).

14. S. B. Hrdy, *The Woman that Never Evolved* (Cambridge, MA: Harvard University Press, 1999).

15. D. Bygott, "Cannibalism among Wild Chimpanzees," *Nature* 238 (1974).

16. M. B. Oliver and J. S. Hyde, "Gender Differences in Sexuality: A Meta-Analysis," *Psychological Bulletin* 114 (1993) (as cited in Geary, Vigil, and Byrd-Craven, "Evolution of Human Mate Choice").

17. G. D. Wilson, "Gender Differences in Sexual Fantasy: An Evolutionary Analysis," *Personality and Individual Differences* 22 (1997) (as cited in Geary, Vigil, and Byrd-Craven, "Evolution of Human Mate Choice").

18. B. J. Ellis and D. Symons, "Sex Differences in Sexual Fantasy: An Evolutionary Psychological Approach," *The Journal of Sex Research* 27 (1990) (as cited in Geary, Vigil, and Byrd-Craven, "Evolution of Human Mate Choice").

19. R. D. Clark III and E. Hatfield, "Gender Differences in Receptivity to Sexual Offers," *Journal of Psychology & Human Sexuality* 2 (1989) (as cited in Geary, Vigil, and Byrd-Craven, "Evolution of Human Mate Choice").

20. L. C. Miller, A. Putcha-Bhagavatula, and W. C. Pedersen, "Men's and Women's Mating Preferences: Distinct Evolutionary Mechanisms?" *Current Directions in Psychological Science* 11 (2002) (as cited in Geary, Vigil, and Byrd-Craven, "Evolution of Human Mate Choice").

21. B. J. Sagarin et al., "Sex Differences (and Similarities) in Jealousy: The Moderating Influence of Infidelity Experience and Sexual Orientation of the Infidelity," *Evolution and Human Behavior* 2 (2003) (as cited in Geary, Vigil, and Byrd-Craven, "Evolution of Human Mate Choice").

22. D. M. Buss et al., "Sex Differences in Jealousy: Evolution, Physiology, and Psychology," *Psychological Science* 3 (1992); D. V. Becker et al., "When the Sexes Need Not Differ: Emotional Responses to the Sexual and Emotional Aspects of Infidelity," *Personal Relationships* 11 (2004); S. M. Murphy et al., "Relationship Experience as a Predictor of Romantic Jealousy," *Personality and Individual Differences* 40 (2006).

23. L. L. Betzig, *Despotism and Differential Reproduction: A Darwinian View of History* (New York: Aldine, 1986).

24. N. A. Chagnon, *Yąnomamö: The Fierce People*, 3rd ed. (New York: Holt, Rinehart & Winston, 1983).

25. R. Hames, "Costs and Benefits of Monogamy and Polygyny for Yąnomamö Women," *Ethology and Sociobiology* 17 (1996).

26. G. Simmons, "How to Sleep with Over 4,800 Women," *British GQ,* March 21, 2012, http://www.gq-magazine.co.uk/comment/articles/2012-03/21/gene-simmons-women-slept-with-dating-tips (accessed October 12, 2013).

27. D. M. Buss, *The Evolution of Desire: Strategies of Human Mating* (New York: Basic Books, 1994).

28. Betzig, *Despotism and Differential Reproduction*.

29. G. P. Murdock and D. R. White, "Standard Cross-Cultural Sample," *Ethnology* 2 (1969).

30. Betzig, *Despotism and Differential Reproduction*, p. 2.

31. Ibid., p. 9.

32. F. G. Poma de Ayala, *The First New Chronical and Good Government* (Paris: Institut d'Ethnologie HRAF Translation, 1936), p. 77 (as cited in Betzig, *Despotism and Differential Reproduction*).

33. Betzig, *Despotism and Differential Reproduction*.

34. Ibid.

35. J. Roscoe, *The Baganda: An Account of their Native Customs and Beliefs* (London: Macmillan, 1911) (as cited in Betzig, *Despotism and Differential Reproduction*).

36. Betzig, *Despotism and Differential Reproduction*, p. 81.

37. W. G. Archer, *The Loves of Krishna in Indian Painting and Poetry* (Mineola: Dover, 2004).

38. E. H. Bryant, *Krishna: A Sourcebook* (New York: Oxford University Press, 2007).

39. T. Gruber, *What the Bible Really Says about Sex: A New Look at Sexual Ethics from a Biblical Perspective* (Worthington, OH: Tom Gruber, 2001).

40. Betzig, *Despotism and Differential Reproduction*.

41. T. W. Doane, *Bible Myths and Their Parallels in Other Religions* (New York: Truth Seeker, 1882).

42. L. Mealey, "The Relationship between Social Status and Biological Success: A Case Study of the Mormon Religious Hierarchy," *Ethology and Sociobiology* 6, no. 4 (1985).

43. B. Nelson, "How Does Power Affect the Powerful," *New York Times*, November 9, 1982, http://www.nytimes.com/1982/11/09/science/how-does-power-affect-the-powerful.html (accessed September 2, 2014).

44. M. C. Langhorne and P. F. Secord, "Variations in Marital Needs with Age, Sex, Marital Status, and Regional Location," *The Journal of Social Psychology* 4 (1955) (as cited in Buss, *Evolution of Desire*, p.27).

45. L. Cosmides and J. Tooby, "Evolutionary Psychology: A Primer," http://www.psych.ucsb.edu/research/cep/primer.html (accessed May 5, 2012).

46. B. Pawlowski and G. Jasienska, "Women's Preferences for Sexual Dimorphism in Height Depend on Menstrual Cycle Phase and Expected Duration of Relationship," *Biological Psychology* 70 (2005).

47. A. C. Little, B. C. Jones, and R. P. Burriss, "Preferences for Masculinity in Male Bodies Change across the Menstrual Cycle," *Hormones and Behavior* 51, no. 5 (2007).

48. I. S. Penton-Voak et al., "Menstrual Cycle Alters Face Preference," *Nature* 399 (1999); I. S. Penton-Voak and D. I. Perrett, "Female Preference for Male Faces Changes Cyclically: Further Evidence," *Evolution & Human Behavior* 21 (2000).

49. S. W. Gangestad et al., "Women's Preferences for Male Behavioral Displays Change across the Menstrual Cycle," *Psychological Science* 15, no. 3 (2004).

50. J. Havlicek, C. S. Roberts, and J. Flegr, "Women's Preference for Dominant Male Odour: Effects of Menstrual Cycle and Relationship Status," *Biological Letters* 1, no. 3 (2005).

51. D. A. Puts et al., "Men's Masculinity and Attractiveness Predict Their Female Partners' Reported Orgasm Frequency and Timing," *Evolution & Human Behavior* 33, no. 1 (2011).

52. R. Thornhill, S. W. Gangestad, and R. Comer, "Human Female Orgasm and Mate Fluctuating Asymmetry," *Animal Behavior* 50 (1995).

53. M. Daly and M. Wilson, "Evolutionary Psychology and Marital Conflict," in *Sex, Power, Conflict: Evolutionary and Feminist Perspectives*, ed. D. M. Buss and N. M. Malamuth (Oxford: Oxford University Press, 1996).

54. D. M. Buss, "Preferences in Human Mate Selection," *Journal of Personality and Social Psychology* 50, no. 3 (1986); D. M. Buss, "Sex Differences in Human Mate Preferences: Evolutionary Hypotheses Tested in 37 Cultures," *Behavioral and Brain Sciences* 12 (1989); E. Hatfield and S. Sprecher, "Men's and Women's Preferences in Marital Partners in the United States, Russia, and Japan," *Journal*

of Cross-Cultural Psychology 26 (1995): 728-50; A. Feingold, "Gender Differences in Mate Selection Preferences: A Test of the Parental Investment Model," *Psychological Bulletin* 112 (1992); M. B. Mulder, "Kipsigis Women's Preferences for Wealthy Men: Evidence for Female Choice in Mammals?" *Behavioral Ecology and Sociobiology* 27 (1990).

55. J. M. Townsend and G. Levy, "Effects of Potential Partners' Costume and Physical Attractiveness on Sexuality and Partner Selection," *Journal of Psychology* 124 (1990).

56. W. G. Graziano, L. A. Jensen-Campbell, and S. G. West, "Dominance, Prosocial Orientation and Female Preferences: Do Nice Guys Really Finish Last?" *Journal of Personality and Social Psychology* 69 (1995).

57. H. L. Harrod, *Renewing the Word: Plains Indian Religion and Morality* (Tucson: University of Arizona Press, 1987).

58. E. A. Peers, Trans., *The Life of St. Theresa of Avila* (London: Sheed and Ward, 1979), pp. 192–93.

59. J. Lafrance, *My Vocation is Love: Therese of Lisieux* (Paris: Mediaspaul, 1994), p. 55.

60. BBC News, "Saudi Police 'Stopped' Fire Rescue," March 15, 2002, http://news.bbc.co.uk/2/hi/1874471.stm (accessed October 22, 2012).

61. A. Bloom and C. Herrman, "The Road North," http://www.pbs.org/frontlineworld/stories/nigeria/thestory.html (accessed October 22, 2012); Associated Press, "Mobs in Nigeria Set Fire to Office, Churches and Bystanders, Killing 50," *Los Angeles Times*, November 22, 2002, http://articles.latimes.com/2002/nov/22/world/fg-nigeria22 (accessed April 4, 2012).

62. Buss, *Evolution of Desire*.

63. Ibid.

64. J. Goodall, *Through a Window: My Thirty Years with the Chimpanzees of Gombe* (New York: Mariner, 2010), p. 65.

65. M. Daly and M. Wilson, "Violence against Stepchildren," *Current Directions in Psychological Science* 5, no. 3 (1996).

66. M. Daly and M. Wilson, "An Assessment of Some Proposed Exceptions to the Phenomenon of Nepotistic Discrimination against Stepchildren," *Annales Zoologici Fennici* 38 (2001).

67. R. A. Gutiérrez, *When Jesus Came, the Corn Mothers Went Away: Marriage, Sexuality and Power in Mexico, 1500–1846* (Stanford: Stanford University Press, 1991).

68. Ibid., p. 76.

69. Ibid., p. 125.

70. Ibid., p. 123.

71. Ibid., p. 314.
72. Ibid., p. 70.
73. Ibid.
74. Ibid., p. 71.
75. Ibid., pp. 126–28.
76. Ibid., p. 131.
77. Ibid., p. 132.
78. Ibid., p. 134.

CHAPTER 5: COOPERATIVE KILLING, IN-GROUP IDENTITY, AND GOD

1. A. Souther, "Warfare Analogy to Virus Infection," http://www.ai.sri.com ~rkf/designdoc/souther-analogy.txt (accessed November 20, 2011).
2. J. K. Choi and S. Bowles, "The Coevolution of Parochial Altruism and War," *Science* 318 (2007).
3. S. Atran, *Talking to the Enemy: Faith, Brotherhood, and the (Un)Making of Terrorists* (New York: Harper Collins, 2010).
4. D. Grossman, *On Killing: The Psychological Cost of Learning to Kill in War and Society* (New York: Bay Back, 1995).
5. D. L. Smith, *The Most Dangerous Animal: Human Nature and the Origins of War* (New York: St. Martin's Griffin, 2007).
6. D. Hume, *A Treatise on Human Nature* (London: Penguin,1985), p. 397 (as cited in Smith, *Most Dangerous Animal*).
7. Smith, *Most Dangerous Animal*, p. 188.
8. H. Sherwood, "The Palestinian Children—Alone and Bewildered—in Israel's Al Jalame Jail," *The Guardian*, January 22, 2012, http://www.guardian.co.uk/world/2012/jan/22/palestinian-children-detained-jail-israel (accessed May 23, 2012).
9. W. Durant and A. Durant, *The Age of Reason Begins: A History of European Civilization in the Period of Shakespeare, Bacon, Montaigne, Rembrandt, Galileo, and Descartes: 1558–1648* (New York: Simon and Schuster, 1961), p 554.
10. Ibid., p. 554.
11. Ibid.
12. J. Vaes, N. A. Heflick, and J. L. Goldenburg, "'We Are People:' Ingroup Humanization as an Existential Defense," *Journal of Personality and Social Psychology* 98, no. 5 (2010).
13. R. Trivers, "The Evolution of Reciprocal Altruism," *The Quarterly Review of Biology* 46 (1971): 35.
14. W. Irons, "Religion as a Hard-to-Fake Sign of Commitment," in *Evolu-*

tion and the Capacity for Commitment, ed. R. M. Nesse (New York: Russell Sage Foundation, 2001), p. 298.

15. R. Sosis, "Religious Behaviors, Badges and Bans: Signaling Theory and the Evolution of Religion," in *Where God and Science Meet*, ed. P. McNamara (Westport, CT: Praeger, 2006).

16. Ibid., pp. 66–7.

17. R. M. Seyfarth and D. L. Cheney, "Grooming, Alliances, and Reciprocal Altruism in Vervet Monkeys," *Nature* 308 (1984); F. de Waal, *Chimpanzee Politics: Power and Sex among Apes* (Baltimore, MD: Johns Hopkins University Press, 1998); F. de Waal, "Food Sharing and Reciprocal Obligations among Chimpanzees," *Journal of Human Evolution* 18 (1989).

18. de Waal, "Food Sharing and Reciprocal Obligations among Chimpanzees."

19. de Waal, *Chimpanzee Politics*.

20. K. R. L. Hall, "Aggression in Monkey and Ape Societies," in *The Natural History of Aggression*, ed. J. Carthy and F. Ebling (London: Academic Press, 1964) (as cited in D. Cummings, "Dominance, Status, and Social Hierarchies," in *The Handbook of Evolutionary Psychology*, ed. D. M. Buss (Hoboken, NJ: Wiley, 2006).

21. R. O. Deaner and A. V. Khera, "Monkeys Pay-Per-View: Adaptive Valuation of Social Images by Rhesus Macaques," *Current Biology* 15 (2005).

22. M. J. Boulton and P. K. Smith, "Affective Bias in Children's Perceptions of Dominance Relationships," *Child Development* 61 (1990) (as cited in Cummings, "Dominance, Status, and Social Hierarchies"); A. E. Russon and B. E. Waite, "Patterns of Dominance and Imitation in an Infant Peer Group," *Ethology & Sociobiology* 2 (1991) (as cited in Cummings, "Dominance, Status, and Social Hierarchies").

23. D. L. Cheney and R. M. Seyfarth, *How Monkeys See the World* (Chicago: University of Chicago Press, 1990); D. Maestripieri, *Macachiavellian Intelligence: How Rhesus Macaques and Humans Have Conquered the World* (Chicago: University of Chicago Press, 2007).

24. T. Nishida, "Alpha Status and Agonistic Alliances in Wild Chimpanzees (Pan troglodytes schweinfurthii)," *Primates* 24 (1983); J. Goodall, *The Chimpanzees of Gombe: Patterns of Behavior* (Cambridge, MA: Harvard University Press, 1986).

25. Cummings, "Dominance, Status, and Social Hierarchies."

26. F. de Waal, "Exploitative and Familiarity-Dependent Support Strategies in a Colony of Semi-Free Living Chimpanzees," *Behavior* 66 (1978).

27. C. Darwin, *The Descent of Man* (Amherst, NY: Prometheus Books, 1998).

28. D. Grossman, *On Killing: The Psychological Cost of Learning to Kill in War and Society* (New York: Bay Back, 1995).

29. S. Junger, *War* (New York: Hachette Book Group, 2010).

30. M. Miller and K. Taube, *An Illustrated Dictionary of the Gods and Symbols of Ancient Mexico and the Maya* (London: Thames and Hudson, 1993).

31. M. Leon-Portilla, *Aztec Thought and Culture* (Norman: University of Oklahoma Press, 1963).

32. Ibid., p. 163.

33. Miller and Taube, *Illustrated Dictionary*.

34. J. Teehan, *In the Name of God: The Evolutionary Origins of Religious Ethics and Violence* (Chichester, UK: Wiley-Blackwell, 2010).

35. Ibid., p. 149.

36. American Psychiatric Association, *Diagnostic and Statistical Manual of Mental Disorders*, 5th ed. (Arlington, VA: American Psychiatric Association, 2013).

37. C. J. Ferguson, "Genetic Contributions to Antisocial Personality and Behavior: A Meta-Analytic Review from an Evolutionary Perspective," *The Journal of Social Psychology* 150, no. 2 (2010).

38. L. Mealey, "The Sociobiology of Sociopathy," in *The Maladapted Mind: Classic Readings in Evolutionary Psychology*, ed. S. Baron-Cohen (East Sussex, UK: Psychology Press, 1997): 169.

39. I. Ishaq, *The Life of Muhammad* (Oxford: Oxford University Press, 1955).

40. V. Bugliosi and C. Gentry, *Helter Skelter: The True Story of the Manson Murders*, 25th anniversary edition (New York: W. W. Norton, 1994).

41. Smith, *Most Dangerous Animal*, p. 100.

42. S. Wells, *Drunk with Blood: God's Killings in the Bible* (SAB Books, 2010).

43. W. Durant and A. Durant, *The Age of Faith: A History of Medieval Civilization—Christian, Islamic, and Judaic—From Constantine to Dante: A. D. 325–1300*, vol. 4 (New York: Simon and Schuster, 1950), p. 592.

44. C. W. Dugger, "Religious Riots Loom over Indian Politics," *New York Times*, July 27, 2002, http://www.nytimes.com/2002/07/27/world/religious-riots-loom-over-indian-politics.html?pagewanted=all&src=pm (accessed October 20, 2011).

45. Wells, *Drunk with Blood*.

46. Ibid.

CHAPTER 6: WHAT IT MEANS TO KNEEL

1. C. Darwin, *The Descent of Man, and Selection in Relation to Sex* (London: John Murray, 1871).

2. C. D. Watkins et al., "Taller Men are Less Sensitive to Cues of Dominance in Other Men," *Behavioral Ecology* 21, no. 5 (2010).

3. C. F. Zink et al., "Know Your Place: Neural Processing of Social Hierarchy in Humans," *Neuron* 58, no. 2 (2008).

4. A. Moors and J. De Houwer, "Automatic Processing of Dominance and Submissiveness," *Experimental Psychology* 52, no. 4 (2005).

5. J. L. Isaac, "Potential Causes and Life-History Consequences of Sexual Size Dimorphism in Mammals," *Mammal Review* 35 (2005); N. Owen-Smith, "Comparative Mortality Rates of Male and Female Kudu: The Costs of Sexual Size Dimorphism," *Journal of Animal Ecology* 62 (1993).

6. S. Pinker, *How the Mind Works* (New York: W. W. Norton, 1997).

7. A. Case and C. Paxson, "Stature and Status: Height, Ability, and Labor Market Outcomes," *Journal of Political Economy* 116, no. 3 (2008).

8. M. Vaz, S. Hunsberger, and B. Diffey, "Prediction Equations for Handgrip Strength in Healthy Indian Male and Female Subjects Encompassing a Wide Age Range," *Annals of Human Biology* 29 (2002).

9. C. von Rueden, M. Guvren, and H. Kaplan, "The Multiple Dimensions of Male Social Status in an Amazonian Society," *Evolution and Human Behavior* 29 (2008).

10. W. E. Hensley, "Height as a Measure of Success in Academe," *Psychology: A Journal of Human Behavior* 30, no. 1 (1993).

11. B. Pawlowski, R. I. M. Dunbar, and A. Lipowicz, "Tall Men Have More Reproductive Success," *Nature* 403 (2000).

12. R. D. Guthrie, *Body Hot Spots: The Anatomy of Human Social Origins and Behavior* (New York: Litton Educational, 1976).

13. B. J. Dixson and P. L. Vasey, "Beards Augment Perceptions of Men's Age, Social Status, and Aggressiveness, but not Attractiveness," *Behavioral Ecology* 23, no. 3 (2012); N. Neave and K. Shields, The Effects of Facial Hair Manipulation on Female Perceptions of Attractiveness, Masculinity, and Dominance in Male Faces," *Personality and Individual Differences* 45 (2008).

14. Dixson and Vasey, "Beards Augment Perceptions."

15. K. B. Starzyk and V. L. Quinsey, "The Relationship between Testosterone and Aggression: A Meta-Analysis," *Aggression and Violent Behavior* 6, no. 6 (2001).

16. P. Schaff and A. C. Coxe, eds., *Nicene and Post-Nicene Fathers*, First Series, vol. 3, *St. Augustine: Expositions on the Psalms* (New York: Cosimo Classics, 2007), p. 623.

17. A. Roberts, J. Donaldson, and A. C. Coxe, eds., *Fathers of the Second Century: Hermas, Tatian, Athenagoras, Theophilus, and Clement of Alexandria* (Buffalo, NY: Christian Literature Publishing Company, 1885); A. Roberts et al., Eds., *Ante-*

Nicene Fathers: Translations of the Fathers Down to AD 325, Volume 2 (Buffalo, NY: Christian Literature Publishing Company, 1894), p. 286.

18. Ibid., p. 275.

19. E. Eckholm and D. Lovering, "Amish Renegades Are Accused in Bizarre Attacks on Their Peers," *New York Times,* October 17, 2011, http://mobile.nytimes.com/2011/10/18/us/hair-cutting-attacks-stir-fear-in-amish-ohio.html (accessed April 24, 2012).

20. L. L. Betzig, *Despotism and Differential Reproduction: A Darwinian View of History* (New York: Aldine, 1986).

21. F. Patrick, *The Primacy of the Apostolic See Vindicated* (Baltimore, MD: Nabu, 1857).

22. W. Burkert, *Creation of the Sacred: Tracks of Biology in Early Religions* (Cambridge, MA: Harvard University Press, 1996).

23. F. de Waal, *Chimpanzee Politics: Power and Sex among Apes* (Baltimore, MD: Johns Hopkins University Press, 1998), p. 78.

24. D. Morris, *The Naked Ape: A Zoologist Study of the Human Animal* (New York: Dell Publishing, 1967), p 146.

25. N. J. Emery, "The Eyes Have It: The Neuroethology, Function, and Evolution of Social Gaze," *Neuroscience and Biobehavioral Reviews* 24 (2000).

26. D. I. Perrett et al., "Social Signals Analyzed at the Single Cell Level: Someone's Looking at Me, Something Touched Me, Something Moved!" *Journal of Comparative Psychology* (1990); D. I. Perrett et al., "Visual Cells in the Temporal Cortex Sensitive to Face View and Gaze Direction," *Proceedings of the Royal Society B* (1985).

27. R. Kawashima et al., "The Human Amygdala Plays an Important Role in Gaze Monitoring: A PET Study," *Brain* 122 (1999).

28. J. C. Gomez, "Ostensive Behavior in Great Apes: The Role of Eye Contact," in *Reaching into Thought: The Mind of Great Apes*, ed. A. E. Russon, K. A. Bard, and S. T. Parker (Cambridge: Cambridge University Press, 1996).

29. Betzig, *Despotism and Differential Reproduction*, p. 31.

30. Betzig, *Despotism and Differential Reproduction*.

31. de Waal, *Chimpanzee Politics*.

32. Homer, *The Odyssey,* trans. R. Fagles (New York: Penguin, 1997).

33. C. I. F. International Association, "Kissing of Hands and Feet of Awliya Allah, Shuyooks and Parents, http://www.cifiaonline.com/kissingofhandsfeet.htm (accessed October 21, 2012).

34. Ibid.

35. N. P. Tanner, *Decrees of the Ecumenical Councils: From Nicea I to Vatican II* (Washington, DC: Georgetown University Press, 1990).

36. H. Chadwick, *Priscillian of Avila: The Occult and the Charismatic in the Early Church* (Oxford: Oxford University Press, 1976).

37. J. M. Anderson, *Daily Life during the Spanish Inquisition* (New Haven, CT: Greenwood, 2002).

38. H. J. D. Denzinger, *Enchiridion Symbolorum: The Sources of Catholic Dogma*, 30th ed., trans. R. J. DeFerrari (St. Louis, MO: B. Herder, 1957).

CHAPTER 7: MALADAPTIVE SUBMISSION TO THE GODHEAD

1. J. Price et al., "The Social Competition Hypothesis of Depression," *British Journal of Psychiatry* 164 (1994).

2. R. C. Kessler et al., "Lifetime Prevalence and Age-of-Onset Distributions of DSM-IV Disorders in the National Comorbidity Survey Replication (NCS-R)," *Archives of General Psychiatry* 62, no. 6 (2005).

3. K. Hodgson and P. McGuffin, "The Genetic Basis of Depression," *Current Topics in Behavioral Neuroscience* 1 (2013).

4. D. R. Wilson, "Evolutionary Epidemiology: Darwinian Theory in the Service of Medicine and Psychiatry," in *Maladapted Mind: Classical Readings in Evolutionary Psychopathology*, ed. S. Baron-Cohen (East Sussex, UK: Psychology Press, 1997), p. 43.

5. T. Schjelderup-Ebbe, "Social Behavior of Birds," in *A Handbook of Social Psychology*, ed. C. A. Murchinson (Worchester, MA: Clark University Press, 1935), p. 966.

6. J. Price, "The Dominance Hierarchy and the Evolution of Mental Illness," *Lancet* 290, no. 7509 (1967).

7. Price et al., "Social Competition Hypothesis of Depression," p. 309–310.

8. O. P. Almeida et al., "Low Free Testosterone Concentration as a Potentially Treatable Cause of Depressive Symptoms in Older Men," *Archives of General Psychiatry* 63, no. 3 (2008).

9. A. Mazur and T. Lamb, "Testosterone, Status, and Mood in Human Males," *Hormones and Behavior* 14 (1980).

10. P. H. Mehta and R. A. Josephs, "Testosterone Change after Losing Predicts the Decision to Compete Again," *Hormones and Behavior* 50 (2006).

11. R. O'Carroll and J. Bancroft, "Testosterone Therapy for Low Sexual Interest and Erectile Dysfunction in Men: A Controlled Study," *British Journal of Psychiatry* 145 (1984).

12. Almeida et al., "Low Free Testosterone Concentration."

13. Mehta and Josephs, "Testosterone Change after Losing."

14. O'Carroll and Bancroft, "Testosterone Therapy."

15. M. McGuire et al., "Dysthymic Disorder, Regulation-Dysregulation Theory, CNS Blood Flow, and CNS Metabolism," in *Subordination and Defeat: An Evolutionary Approach to Mood Disorders and Their Therapy*, ed. L. Sloman and P. Gilbert (New York: Lawrence Erlbaum, 2000).

16. L. A. Kirkpatrick and B. J. Ellis, "The Adaptive Functions of Self-Evaluative Psychological Mechanisms," in *Self-Esteem Issues and Answers: A Sourcebook of Current Perspectives*, ed. M. H. Kernis (New York: Psychology Press; 2006), p. 335.

17. G. A. Parker, "Assessment Strategy and the Evolution of Fighting Behaviour," *Journal of Theoretical Biology* 47, no.1 (1974).

18. P. Gilbert, J. Price, and S. Allan, "Social Comparison, Social Attractiveness, and Evolution: How Might They Be Related?" *New Ideas in Psychology* 13 (1995).

19. J. R. Krebs, N. B. Davies, *An Introduction to Behavioral Ecology*, 3rd ed. (Oxford: Blackwell Scientific Publications, 1993) (as cited in Gilbert, Price, and Allan, "Social Comparison").

20. P. Gilbert and S. Allan, "Assertiveness, Submissive Behaviour and Social Comparison," *British Journal of Clinical Psychology* 33 (1994).

21. C. McFarland and D. T. Miller, "The Framing of Relative Performance Feedback: Seeing the Glass Half Empty or Half Full," *Journal of Personality and Social Psychology* 66, no. 6 (1994).

22. S. Allan and P. Gilbert, "A Social Comparison Scale: Psychometric Properties and Relationship to Psychopathology," *Personality and Individual Differences* 19, no. 3 (1995).

23. American Psychiatric Association, *Diagnostic and Statistical Manual of Mental Disorders*, 5th ed. (Arlington, VA: American Psychiatric Association, 2013), p. 161.

24. J. Price, "Subordination, Self-Esteem, and Depression," in *Subordination and Defeat: An Evolutionary Approach to Mood Disorders and their Therapy*, ed. L. Sloman and P. Gilbert (Mawah, NJ: Lawrence Erlbaum, 2000), p. 172.

25. "Asceticism-Western Asceticism-the Middle Ages," *Science Encyclopedia*, http://science.jrank.org/pages/8388/Asceticism-Western-Asceticism-Middle-Ages.html (accessed May 21, 2012).

26. C. W. Bynum, *Holy Feast and Holy Fast: The Religious Significance of Food to Medieval Women* (Berkeley: University of California Press, 1987), p. 85.

27. Ibid.

28. K. Dervic et al., "Religious Affiliation and Suicide Attempt," *American Journal of Psychiatry* 161, no. 12 (2004).

29. K. Knight, ed., "The Donatists," *Catholic Encyclopedia*, http://www.newadvent.org/cathen/05121a.htm (accessed December 16, 2012).

30. G. D. Chryssides, ed., *Heaven's Gate: Post Modernity and Popular Culture in a Suicide Group* (Burlington, VT: Ashgate, 2011).

31. J. M. Koolhaas et al., "Single Social Defeat in Male Rats Induces a Gradual but Long-Lasting Behavioral Change: A Model of Depression," *Neuroscience Research Communications* 7 (1990).

32. P. Gilbert and M. T. McGuire, "Shame, Status and Social Roles: Psychobiology and Evolution," in *Shame: Interpersonal Behavior, Psychopathology and Culture*, ed. P. Gilbert and B. Andrews (New York: Oxford University Press, 1998).

33. I. S. Bernstein, "Dominance: A Theoretical Perspective for Ethologists," in *Dominance Relations: An Ethological View of Human Conflict and Social Interaction*, ed. D. R. Omark, F. F. Strayer, and D. G. Freedman (New York: Garland, 1980).

34. P. Gilbert, J. Pehl, and S. Allan, "The Phenomenology of Shame and Guilt: An Empirical Investigation," *British Journal of Medical Psychology* 67, no. 1 (2011).

35. Gilbert and McGuire, "Shame, Status and Social Roles."

36. Ibid.

37. W. D. Ray and B. Brown, "Sex and Secularism: The Report," http://ipcpress.com/index.php?id=42 (accessed December 16, 2012).

38. Ibid.

39. C. G. Wilson, "Male Genital Mutilation: An Adaptation to Sexual Conflict," *Evolution and Human Behavior* 29 (2008): 151.

40. R. Wright, *The Evolution of God* (New York: Little, Brown, 2009).

41. M. Miller and K. Taube, *An Illustrated Dictionary of the Gods and Symbols of Ancient Mexico and the Maya* (London: Thames and Hudson, 1993).

42. Ibid.

43. P. Schaff and H. Wallace, eds., *Nicene and Post-Nicene Fathers*, Second Series, vol. 14 (New York: Cosimo Classics, 2007).

44. L. Engelstein, "From Heresy to Harm: Self-Castrators in the Civic Discourse of Late Tsarist Russia," http://src-hokudai-ac.jp/sympo/94summer/chapter1.pdf (accessed November 22, 2012).

45. W. E. A. van Beek et al., "Dogon Restudied: A Field Evaluation of the Work of Marcel Griaule," *Current Anthropology* 32, no. 2 (1991).

46. "Female Genital Mutilation," *World Health Organization*, http://www.who.int/mediacentre/factsheets/fs241/en/index.html (accessed November 23, 2012).

47. A. Hough, "Extramarital Sex 'Causes More Earthquakes', Iranian Cleric Claims," *The Telegraph*, April 19, 2010, http://www.telegraph.co.uk/news/worldnews/middleeast/iran/7606145/Extramarital-sex-causes-more-earthquakes-Iranian-cleric-claims.html (accessed November 20, 2012).

48. M. N. Muller and J. C. Mitani, "Conflict and Cooperation in Wild Chimpanzees," *Advances in the Study of Behavior* 35 (2005).

49. J. C. Mitani and D. Watts, "Demographic Influences on the Hunting Behavior of Chimpanzees," *American Journal of Physical Anthropology* 109 (1999).

50. J. L. Gonzáles, *The Story of Christianity*, vol. 1, *The Early Church to the Dawn of the Reformation* (New York: Harper Collins, 2010).

51. Bynum, *Holy Feast, Holy Fast*, p. 36.

52. "Asceticism-Western Asceticism-the Middle Ages," *Science Encyclopedia*, http://science.jrank.org/pages/8388/Asceticism-Western-Asceticism-Middle-Ages.html (accessed December 16, 2012).

53. Ibid.

54. Ibid.

55. Ibid.

56. American Psychiatric Association, *Diagnostic and Statistical Manual*, p. 161.

57. S. Lemelin and P. Baruch, "Clinical Psychomotor Retardation and Attention in Depression," *Journal of Psychiatric Research* 32, no. 2 (1998): 81–8.

58. T. Yu et al., "Cognitive and Neural Correlates of Depression-Like Behaviour in Socially Defeated Mice: An Animal Model of Depression with Cognitive Dysfunction," *International Journal of Neuropsychopharmacology* 14, no. 3 (2011).

59. P. J. Watson and P. W. Andrews, "Toward a Revised Evolutionary Adaptationist Analysis of Depression: The Social Navigation Hypothesis," *Journal of Affective Disorders* 72 (2002).

60. J. Milton, *Paradise Lost* (New York: Hurd and Houghton, 1869), p. 202.

61. J. Maritain, *The Three Reformers: Luther, Descartes, Rousseau* (Westport, CT: Greenwood, 1970), p. 33.

62. "US Religious Knowledge Survey," *The Pew Forum on Religion and Public Life*, http://www.pewforum.org/uploadedFiles/Topics/Belief_and_Practices/religious-knowledge-full-report.pdf (accessed December 16, 2012).

63. "Gunman Kills Dutch Film Director," *BBC News*, http://news.bbc.co.uk/2/hi/europe/3974179.stm (accessed April 27, 2012).

64. J. Perlez & P. Z. Shaw. "Embassy Attack in Pakistan Kills at Least Six" *New York Times*, http://www.nytimes.com/2008/06/03/world/asia/03pakistan.html (accessed April 27, 2014); L. Polygreen, "Nigeria Counts 100 Deaths Over Danish Caricatures," *New York Times*, February 24, 2006, http://query.nytimes.com/gst/fullpage.html?res=9C06E5DF1F3EF937A15751C0A9609C8B63 (accessed April 27, 2014).

65. "10 Most Censored Countries," *Committee to Protect Journalists*, http://cpj.org/reports/2012/05/10-most-censored-countries.php (accessed June 23, 2012).

66. J. al-Khalili, *The House of Wisdom: How Arabic Science Saved Ancient Knowledge and Gave Us the Renaissance* (New York: Penguin, 2010).

67. S. Atran, "Genesis of Suicide Terrorism," *Science* 299, no. 5612 (2003).

68. "Arab Human Development Report," *United Nations Human Development Programme* (New York: United Nations, 2002) (as cited in S. Harris, *The End of Faith: Religion, Terror and the Future of Reason* (New York: W. W. Norton, 2004)).

69. Harris, *End of Faith*, p. 133.

70. W. Durant and A. Durant, *The Age of Faith: A History of Medieval Civilization–Christian, Islamic, and Judaic–from Constantine to Dante: AD 325-1300*, vol. 4 (New York: Simon and Schuster, 1950), p. 765.

71. Ibid., p.766.

72. Ibid.

CHAPTER 8: THE FEARSOME REPUTATIONS OF APES, MEN, AND GODS

1. J. Goodall, *In the Shadow of Man* (Boston: Houghton Mifflin, 1985).

2. D. Peterson and R. Wrangham, *Demonic Males: Apes and the Origins of Human Violence* (New York: Houghton Mifflin, 1996), p. 191.

3. F. de Waal, *Chimpanzee Politics: Power and Sex among Apes* (Baltimore, MD: Johns Hopkins University Press, 1998).

4. D. Maestripieri, *Macachiavellian Intelligence: How Rhesus Macaques and Humans Have Conquered the World* (Chicago: University of Chicago Press, 2007), p. 72.

5. Ibid.

6. S. Pinker, *The Language Instinct: How the Mind Creates Language* (New York: Harper Collins, 1994).

7. D. L. Smith, *The Most Dangerous Animal: Human Nature and the Origins of War* (New York: St. Martin's Griffin, 2007).

8. "North Korea: Kim Jong-Il's Legacy of Mass Atrocity," *Human Rights Watch*, December 19, 2011, http://www.hrw.org/news/2011/12/19/north-korea-kim-jong-il-s-legacy-mass-atrocity (accessed December 12, 2012).

9. D. Gavlak, "Jordan: 8 Activists Charged for Slandering King," *Associated Press*, September 9, 2012, http://bigstory.ap.org/article/jordan-8-activists-charged-slandering-king (accessed December 12, 2012).

10. L. L. Betzig, *Despotism and Differential Reproduction: A Darwinian View of History* (New York: Aldine, 1986).

11. M. Daly and M. Wilson, *Homicide* (New Brunswick, NJ: Transaction Publishers, 2006), p. 128.

12. D. Cohen et al., "Insult, Aggression, and the Southern Culture of Honor: An 'Experimental Ethnography,'" *Journal of Personality and Social Psychology* 70, no. 5 (1996); R. E. Nisbett and D. Cohen, *Culture of Honor: The Psychology of Violence in the South* (Boulder, CO: Westview, 1996).

13. S. Pinker, *How the Mind Works* (New York: W. W. Norton, 1997).

14. A. V. Papachristos, "Murder by Structure: Dominance Relations and the Social Structure of Gang Homicide," *American Journal of Sociology* 115, no. 1 (2009): 104.

15. Ibid., p. 80.

16. H. W. C. Davis, *The Political Thought of Heinrich von Treitschke* (London: Constable, 1914) (as cited by B. Wyatt-Brown, "The Changing Faces of Honor in National Crises: Civil War, Vietnam, Iraq, and the Southern Factor," *The Johns Hopkins History Seminar, Fall 2005*, p. 1–2, http://www.humiliationstudies.org/documents/WyattBrownFacesHonor.pdf (accessed March 1, 2012)).

17. H. Kissinger, *White House Years* (New York: Simon and Schuster; 1979), p. 288.

18. H. Sidey, *A Very Personal Presidency: Lyndon Johnson in the White House* (New York: Atheneum, 1968) (as cited in Wyatt-Brown, "Changing Faces of Honor").

19. B. Gertz and R. Scarborough, "Shaming Effect on Arab World," *Washington Times*, April 29, 2003 (as cited in Wyatt-Brown, "Changing Faces of Honor," p. 27).

20. Wyatt-Brown, "Changing Faces of Honor."

21. Ibid., p. 30.

22. "Terror in America (30) Retrospective: A Bin Laden Special on Al-Jazeera Two Months before September 11 Bin Laden—The Arab Despair and American Fear," *Middle East Media Research Institute*, December 21, 2001 (as cited by Wyatt-Brown, "Changing Faces of Honor").

23. "Transcript of Osama Bin Laden's October 2001 Interview with Al-Jazeera," *Al-Jazeera*, http://www.againstbush.org/articles/article-16125 1099125987.html (accessed October 30, 2012) (as cited in Wyatt-Brown, "Changing Faces of Honor" p. 35).

24. "Full Text of Bin Laden's 'Letter to America,'" *Observer*, November 24, 2002, http://www.theguardian.com/world/2002/nov/24/theobserver (as cited in Wyatt-Brown, "Changing Faces of Honor," p. 35).

25. R. Dawkins, *The God Delusion* (Boston: Houghton Mifflin, 2006), p. 20.

26. T. Aquinas, *Summa Theologica*, vol. 1, pt. 1, trans. Fathers of the English Dominican Province (New York: Cosimo Classics, 2007).

27. N. C. Lea, *A History of the Inquisition of Spain*, vol. 4 (London: Macmillan, 1906).

28. M. F. Graham, *Blasphemies of Thomas Aikenhead: Boundaries of Belief on the Eve of the Enlightenment* (Edinburgh: Edinburgh University Press, 2008).

29. "A Teddy Bear Nightmare in Sudan," *BBC News*, http://news.bbc.co.uk/2/hi/middle_east/8010407.stm (accessed March 12, 2012).

30. T. Shah and R. Nordland, "Protests Over Koran Burning Reach Kandahar," *New York Times*, April 2, 2011, http://www.nytimes.com/2011/04/03/world/asia/03afghanistan.html (accessed March 20, 2012).

31. A. J. Rubin and G. Bowley, "Koran Burning in Afghanistan Prompts 3 Parallel Inquiries," *New York Times*, February 29, 2012, http://www.nytimes.com/2012/03/01/world/asia/koran-burning-in-afghanistan-prompts-3-parallel-inquiries.html (accessed March 20, 2012).

32. S. Rahimi and A. J. Rubin, "Koran Burning in NATO Error Incites Afghans," *New York Times*, February 21, 2012, http://www.nytimes.com/2012/02/22/world/asia/nato-commander-apologizes-for-koran-disposal-in-afghanistan.html (accessed March 20, 2012).

33. R. Nordland, "In Reactions to Two Incidents, a US-Afghan Disconnect," *New York Times*, March 14, 2012, http://www.nytimes.com/2012/03/15/world/asia/disconnect-clear-in-us-bafflement-over-2-afghan-responses.html?pagewanted=all (accessed March 30, 2012).

34. A. Coulter, "This Is War: We Should Invade Their Countries," *National Review*, September 13, 2001. http://old.nationalreview.com/coulter/coulter.shtml (accessed March 15, 2012).

35. F. de Waal, *Chimpanzee Politics*.

CHAPTER 9: GOD'S TERRITORY

1. B. Diaz del Castillo, *The Discovery and Conquest of Mexico* (New York: Farrar, Straus, and Giroux, 1956).

2. B. Macintyre, "The Dignified Reply to the War-Grave Vandals," *Times*, March 6, 2012, http://www.thetimes.co.uk/tto/opinion/columnists/benmacintyre/article3341203.ece (accessed December 1, 2012).

3. "Ansar Dine Destroy more Shrines in Mali," *Al Jazeera*, July 10, 2012, http://www.aljazeera.com/news/africa/2012/07/201271012301347496.html (accessed August 20, 2013).

4. I. Singleton and C. P. van Schaik, "The Social Organization of a Population of Sumatran Orangutans," *Folia Primatol* 73 (2002).

5. B. Galdikas, "Orangutan Reproduction in the Wild," in *Reproductive Biology of the Great Apes*, ed. C. E. Graham (New York: Academic Press, 1981), p. 288.

6. J. C. Mitani, "Mating Behaviour of Male Orangutans in the Kutai Game Reserve, Indonesia," *Animal Behaviour* 33 (1985).

7. J. C. Mitani, D. P. Watts, and S. J. Amsler, "Lethal Intergroup Aggression Leads to Territorial Expansion in Wild Chimpanzees," *Current Biology* 20, no. 12 (2010); M. N. Muller and J. C. Mitani, "Conflict and Cooperation in Wild Chimpanzees," *Advances in the Study of Behavior* 35 (2005).

8. J. M. Williams, *Female Strategies and the Reasons for Territoriality in Chimpanzees: Lessons from Three Decades of Research at Gombe* (Minneapolis: University of Minnesota, 1999) (as cited in D. Watts and J. C. Mitani, "Boundary Patrols and Intergroup Encounters in Wild Chimpanzees," *Behaviour* 138 [2001]).

9. T. Nishida, M. Hiraiwa-Hasegawa, and Y. Takahata, "Group Extinction and Female Transfer in Wild Chimpanzees in the Mahale National Park, Tanzania," *Zeitschrift für Tierpsychologie* 67 (1985).

10. K. Wolf and S. R. Schulman, "Male Response to 'Stranger' Females as a Function of Female Reproductive Value," *American Naturalist* 123 (1984).

11. M. L. Wilson and R. W. Wrangham, "Intergroup Relations in Chimpanzees," *Annual Review of Anthropology* 32 (2003).

12. J. Goodall et al., "Inter-Community Interactions in the Chimpanzee Populations of the Gombe National Park," in *The Great Apes,* ed. D. Hamburg and E. McCown (Menlo Park, CA: Benjamin/Cummings, 1979); J. Goodall, *The Chimpanzees of Gombe: Patterns of Behavior* (Cambridge, MA: Harvard University Press, 1986).

13. J. Williams, G. Oehlert, and A. Pusey, "Why do Male Chimpanzees Defend a Group Range?" *Animal Behaviour* 68 (2004).

14. Watts and Mitani, "Boundary Patrols and Intergroup Encounters."

15. Ibid.

16. W. Rodzinski, *A History of China* (Oxford: Pergamon, 1979), p. 165 (as cited in D. L. Smith, *The Most Dangerous Animal: Human Nature and the Origins of War* (New York: St Martin's Griffin, 2007)).

17. P. Schrijvers, *The GI War against Japan: American Soldiers in Asia and the Pacific during World War II* (New York: New York University Press, 2002).

18. I. Chang, *The Rape of Nanking: The Forgotten Holocaust of World War II* (New York: Penguin, 1997).

19. A. Beevor, *The Fall of Berlin 1945* (New York: Penguin, 2003).

20. J. Hatzfeld, *Machete Season: The Killers in Rwanda Speak* (New York: Farrar, Straus, and Giroux, 2005) (as cited in Smith, *Most Dangerous Animal*).

21. P. A. Weitsman, "The Politics of Identity and Sexual Violence: A Review of Bosnia and Rwanda," *Human Rights Quarterly* (2008).

22. W. Durant and A. Durant, *The Age of Reason Begins: A History of European Civilization in the Period of Shakespeare, Bacon, Montaigne, Rembrandt, Galileo, and Descartes: 1558–1648* (New York: Simon and Schuster, 1961), p. 527.

23. Smith, *Most Dangerous Animal.*

24. A. Grossman, "Single Islamic State Militant 'has Killed 150 Women and Girls' because They Refused to Marry Members of the Terrorist Group," Daily Mail, December 18, 2014, http://www.dailymail.co.uk/news/article-2878693/Single-Islamic-State-militant-killed-150-women-girls-refused-marry-members-terrorist-group.html (accessed December 20, 2014).

25. I. Watson, "Treated like Cattle: Yazidi Women Sold, Raped, Enslaved by ISIS," CNN, November 7, 2014, http://www.cnn.com/2014/10/30/world/meast/isis-female-slaves/ (accessed December 20, 2014).

26. "Boko Haram Insurgents Kill 100 People as They Take Control of Nigerian Town," Guardian, July 19, 2014, http://www.theguardian.com/world/2014/jul/19/boko-haram-kill-100-people-take-control-nigerian-town (accessed December 20, 2014).

27. D. Suzuki, *The Legacy: An Elder's Vision for Our Sustainable Future* (Vancouver, BC: Greystone, 2010), p. 23.

28. T. Malthus, *An Essay on the Principles of Population* (Oxford: Oxford University Press, 2008).

29. Science Summit on World Population, "A Joint Statement by the 58 of the World's Scientific Academies," *Population and Development Review* 20 (1994) (as cited in Suzuki, *Legacy*, p. 20).

30. Malthus, *Essay on the Principle of Population*, p. 61.

31. D. M. Buss, "Sex Differences in Human Mate Preferences: Evolutionary Hypotheses Tested in 37 Cultures," *Behavioral and Brain Sciences* 12 (1989).

32. V. Griskevicius et al., "The Financial Consequences of Too Many Men: Sex Ratio Effects on Saving, Borrowing, and Spending," *Journal of Personality and Social Psychology* 102, no. 1 (2012).

33. Ibid.

34. B. W. Husted, "Culture and Ecology: A Cross-National Study of the Determinants of Environmental Sustainability," *Management International Review* 45, no. 3 (2005).

35. Ibid.

36. G. Hofstede, *Culture and Organization: Software of the Mind* (New York: McGraw Hill, 1997), p. 28 (as cited by Husted, "Culture and Ecology").

37. G. Hofstede, "Dimensionalizing Cultures: The Hofstede Model in Context," *Online Readings in Psychology and Culture* Unit 2 (2011).

38. Hofstede, "Dimensionalizing Cultures."

39. Husted, "Culture and Ecology."

40. H. Park, C. Russell, and J. Lee, "National Culture and Environmental Sustainability: A Cross-National Analysis," *Journal of Economics and Finance* 31, no. 1 (2007).

41. Ibid.
42. Ibid., pp. 113–14.
43. E. F. Kittay, "Woman as Metaphor," *Hypatia* 3, no. 4 (1988): 1.
44. P. Horgan, *Conquistadors in North American History* (El Paso: Texas Western, 1982), pp. 225–26.
45. S. Brooks-Thistlethwaite, "Women's Religious Freedom Violated: Photo of All Male Birth Control Witnesses Tells the Viral Truth," *Washington Post*, February 16, 2012, http://www.washingtonpost.com/blogs/guest-voices/post/womens-religious-freedom-violated-photo-of-all-male-birth-control-witnesses-tells-the-viral-truth/2012/02/16/gIQAeyykIR_blog.html (accessed March 12, 2012); L. Bassett and A. Terkel, "House Democrats Walk Out of One-Sided Hearing on Contraception, Calling It an 'Autocratic Regime,'" *Huffington Post*, February 16, 2012, http://www.huffingtonpost.com/2012/02/16/contraception-hearing-house-democrats-walk-out_n_1281730.html (accessed March 12, 2012).
46. D. Suzuki, "The Legacy of David Suzuki," *PRI's Environmental News Magazine*, December 17, 2010, http://www.loe.org/shows/segments.html?programID=10-P13-00051&segmentID=6 (accessed April 23, 2013).

CHAPTER 10: RIGHTING OURSELVES

1. L. Sherr, *Failure Is Impossible: Susan B. Anthony in Her Own Words* (New York: Times Books, 1995), p. 255.
2. J. Goodall, "Gombe Chimpanzee Politics," *Primate Politics*, ed. G. A. Schubert and R. D. Masters (Carbondale, IL: Southern Illinois University Press, 1991), p. 137.
3. Plato, *The Republic*, trans. Robin Waterfield (Oxford: Oxford University Press, 1993).
4. R. Sosis, "Religion and Intragroup Cooperation: Preliminary Results of a Comparative Analysis of Utopian Communities," *Cross-Cultural Research* 34, no. 1 (2000).
5. C. Darwin, *The Descent of Man* (Amherst, NY: Prometheus Books, 1998).
6. R. Wright, *The Moral Animal: Why We Are the Way We Are: The New Science of Evolutionary Psychology* (New York: Vintage, 1995), p. 375.
7. P. Norris and R. Inglehart, *Sacred and Secular: Religion and Politics Worldwide* (Cambridge, UK: Cambridge University Press, 2004).
8. L. Mealey, "The Sociobiology of Sociopathy," in *The Maladapted Mind: Classic Readings in Evolutionary Psychology*, ed. S. Baron-Cohen (East Sussex, UK: Psychology Press, 1997).

9. R. Wright, "Why Can't We All Just Get Along? The Uncertain Biological Basis of Morality," *The Atlantic*, http://www.theatlantic.com/magazine/archive/2013/11/why-we-fightand-can-we-stop/309525/ (accessed December 3, 2012).

10. P. Demieville, "Buddhism and War," in *Buddhist Warfare*, ed. M. Jerryson and M. Juergensmeyer (Oxford: Oxford University Press, 2010).

11. Ibid.

12. Ibid.

13. Ibid.

14. Ibid., p. 42.

15. S. Jenkins, "Making Merit through Warfare and Torture According to the Ārya-Bodhisattva-gocara-upāyaviṣaya-vikurvaṇa-nirdeśa Sūtra," in *Buddhist Warfare*, ed. M. Jerryson and M. Juergensmeyer (Oxford: Oxford University Press, 2010), p. 68.

16. B. Faure, "Afterthoughts," in *Buddhist Warfare*, ed. M. Jerryson and M. Juergensmeyer (Oxford: Oxford University Press, 2010).

17. Dalai Lama, "Nobel Lecture, December 11, 1989," *Nobelprize.org*, http://www.nobelprize.org/nobel_prizes/peace/laureates/1989/lama-lecture.html (accessed June 23, 2014).

18. The Buddha, *The Dhammapada*, trans. G. Fronsdal (Boston: Shambhala, 2006), p. 35.

19. Ibid., p. 36.

20. S. Harris, *Letter to a Christian Nation* (New York: Random House, 2006), p. 21.

21. It is uncertain whether these are the words spoken by the Buddha, but they, like many other proverbs, have become central tenets of Buddhism.

22. T. Jefferson, *The Jefferson Bible: What Thomas Jefferson Selected as the Life and Morals of Jesus of Nazareth* (Thousand Oaks, CA: Lakewood, 2011), p. 9.

23. Wright, "Why Can't We All Just Get Along?"

24. J. Greene, *Moral Tribes: Emotion, Reason, and the Gap Between Us and Them* (New York: Penguin, 2013).

25. N. Guetin, *Religious Ideology in American Politics: A History* (Jefferson, NC: McFarland, 2009).

26. C. H. Moehlman, *The American Constitutions and Religion. Religious References in the Thirteen Colonies and the Constitutions of the Forty-Eight States: A Sourcebook of Church and State in the United States* (Clark, NJ: The Lawbook Exchange, 2007), p. 38.

27. R. Dawkins, *The God Delusion* (Boston: Houghton Mifflin, 2006).

28. R. Ahdar and I. Leigh, *Religious Freedom in the Liberal State*, 2nd ed. (Oxford: Oxford University Press, 2013), p. 70.

29. Dawkins, *God Delusion*, p. 62.

30. Norris and Inglehart, *Sacred and Secular*.

31. "Lobbying for the Faithful," *Pew Research Center*, http://www.pewforum

.org/2011/11/21/lobbying-for-the-faithful-exec/#expenditures (accessed November 20, 2013).

32. T. Riley, "Messing with Texas Textbooks," *Moyers & Company*, http://billmoyers.com/content/messing-with-texas-textbooks/ (accessed November 18, 2013).

33. J. Calvin, "A Harmony of the Gospels Matthew, Mark and Luke Volume III, and the Epistles of James and Jude," in *Calvin's Commentaries*, trans. A. W. Morrison, ed. D. W. Torrance and T. F. Torrance (Grand Rapids, MI: The Saint Andrews Press, 1972), p. 245.

34. Riley, *Messing with Texas Textbooks*.

35. S. Harris, *The Moral Landscape: How Science Can Determine Human Values* (New York: Free Press, 2010).

36. Plato, *Apology*, trans. B. Jowett, ed. J. Manis, An Electronic Classics Series Publication, Pennsylvania State University, Hazelton, PA, http://www2.hn.psu.edu/faculty/jmanis/plato/apology.pdf (accessed December 20, 2013).

37. P. Zuckerman, *Society Without God* (New York: New York University Press, 2008) (as cited in Harris, *Moral Landscape*, p. 146).

38. Norris and Inglehart, *Sacred and Secular*; G. Paul, "The Chronic Dependence of Popular Religiosity upon Dysfunctional Psycholosociological Conditions," *Evolutionary Psychology* 73, no. 3 (2009).

39. M. Daly, M. Wilson, and S. Vasdev, "Income Inequality and Homicide Rates in Canada and the United States," *Canadian Journal of Criminology* 43, no. 2 (2001).

40. Norris and Inglehart, *Sacred and Secular*.

41. Harris, *Letter to a Christian Nation*.

42. D. L. Hall, D. C. Matz, and W. Wood, "Why Don't We Practice What We Preach? A Meta-Analytic Review of Religious Racism," *Personality and Social Psychology Review* 14, no. 10 (2010).

43. Norris and Inglehart, *Sacred and Secular*.

44. Paul, "Chronic Dependence of Popular Religiosity," p. 25.

45. Ibid.

46. F.-A. Aulard, *The French Revolution: A Political History, 1789-1804* (New York: Scribner, 1917), p. 205.

47. D. M. Buss, *The Evolution of Desire: Strategies of Human Mating* (New York: Basic Books, 1994), p. 69.

INDEX

!Kung San, 19

Achlerin, Elsbeth, 169
Adad, 21
Adams, John, 234
Aikenhead, Thomas, 190–91
Alexander the Great, 180
Alexander VI (pope), 59
Alfonso the Battler, 180
Alfonso the Brave, 181
Alfonso the Conqueror, 181
Al-Ma'mun, 173
Amenhotep IV, 21–22
American football, 205
Amun, 21–22
Ansar Dine, 198
Anthony, Susan B., 223
antisocial personality, 120–21, 123–24, 127, 129, 230
Anu, 20
Astarte, 23, 190
Aten, 21–22
Augustine, Saint, 54, 55, 133
Azande, 71

Baal, 23, 190, 200
baboons, 30, 46, 52
Banu Qurayza, battle of the, 210
beards, 108, 127, 133–34, 184
Benedict VI (pope), 59
Benedict XVI (pope), 53
Benevenuta of Bojanis, 169
Bes, 46–47
Bin Laden, Osama, 185

blasphemy, 47, 76, 82, 187–91, 194, 237
Boko Haram, 211
Bonaparte, Napoleon, 60, 247
bonobos, 15–16, 66
book burning, 147, 172, 236
Bosnian War, 206–7
Boykin, William, 61
Buddhism
    the Buddha, 136, 231
    Buddhas of Bamiyan, 197
    heresy in, 231
    objectivity in, 233–34
    pacifism and, 230–31
    regalia, 135
    sexual conquest in, 232
    warfare and, 231–32
Bundy, Ted, 123
Burr, Aaron, 183
Bush, George W., 49, 56, 60–61, 125, 185

Caesar, Julius, 177
Catherine of Siena, Saint, 169
Ceaușescu, Nicolae, 177
Chamorro, Violeta, 50
Charlemagne (king), 181
Chemosh, 23, 190
chimpanzees (robust)
    affiliation/alliance, 56, 109
    attacking genitals, 66
    dominance hierarchies, 16
    food resources, 165
    infanticide, 67

in-group aggression, 16, 30, 57, 64, 88, 109, 136–37, 165, 178, 193
   keeping alphas in check, 193
   posturing/agonistic displays, 16, 136–37, 178
   providing protection, 30, 46
   punishing female infidelity, 88
   sexual dominion, 64–65
   warfare, 204–5, 224
Christopher (pope), 59
cloistering, 35, 72, 95
Cojuangco-Aquino, Maria Corazon, 49
Commission for the Promotion of Virtue and Prevention of Vice, 87
Conquista (Spanish), 92–98, 147, 196–97, 219–20
Costly Signaling Theory, 108–9, 110, 111–16, 128
Coulter, Ann, 192–93
Crusades, 125–26, 174, 185, 192, 231
cuckoldry, 34–35, 53–54
Cushan-Rishathaim (king), 24

Dahomey, kings of, 71–72
Dalai Lama, 135
Darwin, Charles, 27–28, 29, 45–46, 111, 226
depression
   cognitive impairment, 169–70
   chickens and, 152
   lifetime prevalence, 151
   macaques and, 152
   self-harm, 157–58
   social competition hypothesis, 152–53
   suicide, 158
   testosterone and, 153
   worthlessness, 154
Devaki, 78
divine right of kings, 145
Dogon tribe, 164
dominance versus prestige, 16

ecofeminism, 217
Eglon (king), 24
Enlil, 20
evolutionary psychology of religion, 41–42
eye contact, 16, 17, 137, 140–41, 149, 155, 156, 160

facial resemblance, 54, 56
faith-based charities, 245, 246
falsifiability, 242
fictive kin, 101
fields of Aaru, 52
First Vatican Council, 145
Foundation Stone, 201
Francis of Assisi, Saint, 169
Francis of Bonaventure, Saint, 169
Fur tribe, 140

Gadhafi, Moammar, 177
Gampo, Songtsan, 232
Ganda, kings of, 72
Gaona, 19
Genghis Khan, 206
genital mutilation
   castration, 59, 66, 67, 71, 82–83, 159, 163, 164
   circumcision, 108, 112, 162–63
   cliterodectomy, 11, 164
   in Mesoamerica, 111, 163
   penectomy, 92–94
gluttony, 159, 166
gorillas, 64, 66
Gregory VII (pope), 142–43
Gujarat, India massacre, 127

Hades, 52
Hagee, John, 60
halos, evolutionary significance of, 135–36
Hamilton, Alexander, 183
Hammurabi, 20
heaven, 48, 52, 54, 55, 75, 78, 79, 83, 85, 86, 138, 139, 141, 174, 201, 219, 227, 233

hell, 52, 115, 138, 157, 161, 174, 199, 231–32
Hera, 73
Herakles, 47
heresy, 145, 146, 173, 231, 234
Herod (king), 157
Hinduism, 52, 73, 74, 127, 133, 227, 232, 233
Holy Sepulchre, 201
honor cultures, 182–84
House of Knowledge in Baghdad, 173
Huitzilopochtli, 112–13
Hussein, Saddam, 177, 184
Hyperactive Agency Detection Device (HADD), 42

imago Dei, 54–55, 56, 58, 218, 222, 229
Index Librorum Prohibitorum, 172
indirect reciprocity, 107, 108
indulgences, 174–75
infanticide
   in gods, 56, 91–92, 188, 208, 223, 229
   in men, 56, 90–92, 105
   in non-human primates, 52, 56, 66–67
infrahumanization, 102–5, 147, 182
Innocent III (pope), 146
Inquisition, 11, 146, 149, 189, 234
Islamic State (ISIS), 211
Islamist Movement for Oneness and Jihad in West Africa (MOJWA), 198
Ismail the Bloodthirsty, 180, 181
Issa, Darrell, 221
Ivan the Terrible, 180

Jackson, Andrew, 183
James, William, 39, 98, 223
James the Conqueror, 181
Jefferson, Thomas, 234, 235, 237–38, 239, 240
Jeffersonian Bible, 234
Johnson, Lyndon B., 184
John XI (pope), 59
John XII (pope), 58–59

John XIV (pope), 59
Jokhang Temple in Lhasa, 232
Josiah (king), 23–24
Jyllands-Posten, 172–73

Kafa tribe, 134
Kim Jong Il, 56, 181, 189
Kim Jong Un, 56
kin altruism 37–38, 52–56, 101
Kish, king of, 20
Kissinger, Henry, 83, 184
Krishna, 73–74, 76, 78

Lent, 75, 168
Leo V (pope), 59
Leo VIII (pope), 59
Lidwina of Schiedam, Saint, 169
Luther, Martin, 171, 237

macaques
   alliances in, 109
   attacking genitalia, 66
   attraction to high status, 109
   depression among, 153
   line warfare, 179
   mating behaviors in, 33, 65
   punishing female infidelity, 65
Madison, James, 238
male spending behavior, 215–16
Malthus, Thomas, 213–14, 221
Manifest Destiny, 221
Manson, Charles, 123
Marcos, Ferdinand, 49
Marduk, 20, 21, 24
Masau'u, 52
mate selection, 30
metamorality, 235
Milcom, 23, 190
mind reading, 41
mixed martial arts, 205
monolatry, 21
Morning Star (Blackfoot god), 84
Motecuhzoma, 113

Nabu, 21
Native Americans
    Aztec, 58, 111, 112–13, 163, 197, 220
    Blackfoot, 84
    First Nations, 222
    Hopi, 52
    Inca, 58, 71, 72, 78, 220
    Klamath, 19
    Maya, 58, 111, 113, 163, 220
    Puebloan, 93–97, 219
    Yąnomamö, 29, 70
    Zapotec, 163
Nebuchadnezzar (king), 24
Nicholas of Flüe, Saint, 169
Nicolas the Bloody, 181
Nilus the Elder, Saint, 166
9/11, 49, 185, 192

Obama, Barack, 221
Odin, 52
Oñate, Juan de, 93, 219
orangutans, 133, 202–3
Orthodox Christians, 134, 142
Osiris, 52
overpopulation, 213–15, 219, 229–30

papal infallibility, 145
papal tiara, 135
paradise, 52, 78, 115–16, 227
parochial altruism, 100
patas monkeys, 46
Paul VI (pope), 172
Pew Research Center, 172, 239
pharaohs, 21, 58, 60, 91, 134
Plato, 224
Priscillian, 146

Quetzalcoatl, 111

Ram, 127
Ramadan, 75, 168
Rape of Nanking, 206
rape
    God ordering, 89, 207–9

    gods committing, 19, 73
    men protecting from, 31
    in orangutans, 203
    priests committing, 94
    punishment for women in Islam, 89
    wartime and, 93, 105, 126, 127, 184–85, 206–7, 211
Rayhana, 210
reciprocal altruism, 107, 108, 114, 193
religious fasting, 146, 167–69, 188
religious lobby, 239
religious mortification, 157–58
resource holding potential, 154–55, 170
retribution theology, 24–25
Ripoll, Cayetano, 146
Rudra, 232

Satan, 47–48, 51–52, 60, 61, 128, 136, 138, 171, 189, 190, 199, 237
Science Summit on World Populations, 213
Scipio Africanus, 181
selfish gene, 38
Selim the Brave, 181
separation between church and state, 234, 235, 239–40
Sergius III (pope), 59
sex in gods, 19, 20, 73, 80, 95–96
sexual shame, 11, 95, 160–61
Sharia Law, 90, 191, 197
Sirleaf, Ellen Johnson, 50
Skoptsy, 163–64
Societal health research, 243–44
Socrates, 242
Solomon (king), 74
Srinmo, 232
Stalin, Joseph, 22, 236
Suradji, Ahmad, 123

Taliban, 147, 197, 198, 234
teddy bear blasphemy case, 191
temporal lobe epilepsy, 40
Tenochtitlán, 197

Teresa of Ávila, Saint, 84–85, 169
Terror Management Theory, 48–51, 105
Tertullian, 165–66
testosterone
    aggression and, 50, 133
    competition and, 153
    facial features and, 50
    facial hair and, 133
    mood and, 153
    sex drive and, 153
Texas State Board of Education, 239–40
Thérèse of Lisieux, Saint, 169
Thirty Years' War, 104–5, 207
Thomas Aquinas, Saint, 54, 171, 189, 191
three Bs, 108
Thunder (Blackfoot god), 84
Tikopians, 140
Tohil, 111
Tomb of King David, 202
Tongans, 19
Trudell, John, 195

Umma, king of, 20
underworld, 52

Valhalla, 52
Van Gogh, Theodoor, 172
veiling, 35, 87–88, 108, 164

Virginia Statute for Religious Freedom, 237–38
virgins
    God kills, 89, 114
    God spares, 82
    gods prefer, 76–80
    hymen repair, 35
    prizes of powerful men, 72, 135
    among men, 163
    in paradise, 210, 227
    as spoils of Biblical war, 91, 208, 209
Vishnu, 79
Vlad the Impaler, 180

Wailing Wall, 201
William the Conqueror, 181
Wolfowitz, Paul, 184

Xenophanes, 11

Yahweh, 22, 23, 24, 25, 74, 190, 197, 200

Zande, 71
Zedekiah (king), 24
Zeus, 47, 58, 73, 76